D0045709

ALSO BY GIULIO TONONI

A Universe of Consciousness: How Matter Becomes Imagination

(with Gerald Edelman)

PHI

Φ

This Book is a Gift of:

FRIENDS OF THE
LAFAYETTE LIBRARY

Contra Costa County Library

WITHDRAWN

PHI

A Voyage from the Brain to the Soul

GIULIO TONONI

PANTHEON BOOKS · NEW YORK

Copyright © 2012 by Giulio Tononi

All rights reserved. Published in the United States by Pantheon
Books, a division of Random House, Inc., New York, and in
Canada by Random House of Canada Limited, Toronto.

Pantheon Books and colophon are registered trademarks
of Random House, Inc.

Grateful acknowledgment is made to Alfred A. Knopf for
permission to reprint "b o d y" copyright © 1995 by James Merrill,
from *A Scattering of Salts* by James Merrill. Used by permission of
Alfred A. Knopf, a division of Random House, Inc.

LIBRARY OF CONGRESS CATALOGING-IN-PUBLICATION DATA
Tononi, Giulio.
Phi : a voyage from the brain to the soul / Giulio Tononi.
p.   cm.
ISBN 978-0-307-90721-9 (hardback)
1. Consciousness—Physiological aspects.   2. Brain—
physiology.   3. Mind and body.   I. Title.
QP411.T66 2012   612.8—dc23   2011041620

www.pantheonbooks.com

Jacket design by Giulio Tononi

Printed in Singapore
First Edition
2 4 6 8 9 7 5 3 1

CONTENTS

6

A BRAIN LOCKED IN

*In which is shown that motor outputs and pathways
are not necessary for consciousness,
nor are they sufficient*

7

EMPRESS WITHOUT MEMORY

*In which is shown that many brain circuits
that help us see, hear, remember, speak, and act
are not necessary for consciousness*

8

A BRAIN SPLIT

*In which is shown that consciousness is divided
if the brain is split*

9

A BRAIN CONFLICTED

*In which is said that consciousness
can split if different regions of the brain
refuse to talk to each other*

10

A BRAIN POSSESSED

*In which is shown that when cortical neurons
fire strongly and synchronously, as during certain seizures,
consciousness fades*

11

A BRAIN ASLEEP

*In which is shown that when cortical neurons
can be on and off only together, as during dreamless sleep,
consciousness fades*

· PART II ·

31

DAYLIGHT III: CONSCIOUSNESS GROWING
In which is said that, by growing consciousness,
the universe comes more into being,
the synthesis of one and many

32

EPILOGUE

33

AFTERTHOUGHTS

PREFACE

Every night, when we fall into dreamless sleep, consciousness fades. With it fades everyone's private universe—people and objects, colors and sounds, pleasures and pains, thoughts and feelings, even our own selves dissolve—until we awake, or until we dream.

What is consciousness, and what does it mean? How is it related to the world around us? What is it made of, and how is it generated inside the brain? Can science shed some light on it? Perhaps, but consciousness cannot just rest inside the shroud of science. Because consciousness is more than an object of science: it is its subject too.

What follows is a story where an old scientist, Galileo, goes through a journey in search of consciousness. In his time, Galileo removed the observer from nature and opened the way for the objectivity of science. Perhaps this is why Galileo is engaged to return the observer to nature, to make subjectivity a part of science. Or perhaps because Galileo was a master of thought experiments, of which this book makes much use.

During his journey, Galileo meets people from his and other times, learns many lessons, thinks many thoughts, and sometimes wonders, too, whether he is awake or dreaming. But each chapter makes some kind of statement, building on the previous ones, and Galileo's understanding grows. So in the first part of the book, he learns the facts of consciousness and the brain—why certain parts of the brain are important but not others, or why consciousness fades with sleep. In the second part, he sees how these facts can be unified and understood through a scientific theory of consciousness—a theory that links consciousness to Φ, the symbol of integrated information that gives the book its title. And finally, in the third part of the book, he realizes some of the theory's implications, and sees that they concern us all, because consciousness is everything we have, and everything we are. Each experience, Galileo realizes, is a unique shape made of integrated information—a shape that is maximally irreducible—the shape of understanding. And it is the only shape that's *really* real—the most real thing there is. The reader can judge whether the old man's musings make any sense at all.

The notes at the end of each chapter attempt to clarify some ingredients of the main text and list credits when they could be identified. (Some pictures and quotes were liberally altered.) Those interested in a scientific exposition of a theory similar to the one presented in the main text can consult "Consciousness as Integrated Information," *Biological Bulletin* (2008), and references therein. Several thoughts, images, and citations have appeared in previous work—it would seem that some people cannot help but write the same story all their life.

PROLOGUE

The Dream of Galileo

Midway upon the journey of his dream,
he found himself adrift inside the dark,
not knowing whence or wherefore he was there.

He looked and saw that all was black. His soul was rising fast, or it was falling—Galileo knew not which way was up. So his soul turned and saw the stars. He saw the galaxies, fixed in their distant ways, and saw the planets, revolving in the indifferent void. The earth was moving, too, but there was no sun.

Yet dawn was coming. The earth painted itself like a faint half-moon, one face toward sunrise and one toward sunset. It loomed closer and closer—he could see the soft light of the morning sweep the ridges of the highest mountains, the shadows recede over the valleys. Soon the top of the tallest trees stared into the light, and amid the woods the monastery of his youth appeared—his own small turf of earth had come to meet its guest at his own place.

Hovering in midair, he slid inside his erstwhile room—a soul floating above his bed—and the soul saw himself: his eyes were closed, his mouth half open—it was an old man's face. And yet the soul was light, unfastened from the frame of his own body.

His b-o-d-y! He could see its letters,

> entering (stage right), then floating full,
> then heading off—so soon—
> how like a little kohl-rimmed moon
> o plots her course from b to d
> —as y, unanswered, knocks at the stage door.

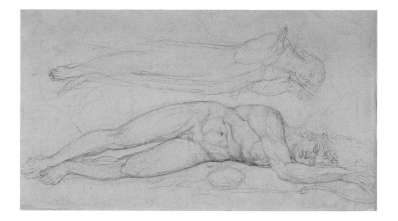

He put his ear to his own chest and listened to the heart. How could the pulse go on, beat after beat, for all of life? No machine could run that long without a stumble. Ask not if the beating cranks are going to jam, he thought, but when.

He heard his breath flow in and out. So for a while he watched the bellows puff and wheeze, in and out, and wondered how many puffs were left in him. Ask not, he thought, for far too soon the steam is going to fizzle—every balloon leaks.

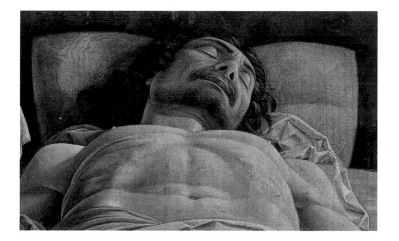

Then, suddenly, straining to blow inside his body, the air took Galileo with it. And he felt his soul being sucked inward, flowing through narrow nostrils inside the dark vault of the skull.

Within he saw another sky: the black sky of his skull—there were no stars in it. But his soul turned again and saw another earth. Floating over its vast continents, he saw its mountains and soared above its valleys. Another dawn was rising: the ridges and the valleys were lined with forests, perhaps they were the trees of youth, the shady valley where he had once learned what nature was and what the soul was not. The smell of youth wafting through the air, the rounded mountains were breathing, too, beating slowly and softly with his pulse. And he saw then what it was, the expanse below his soul: it was the brain, the brain that shone with light as if it held its sun inside.

Then he landed and dived into the brain, amongst its trees, under the canopy, and deep inside the brain he saw his monastery again, and the long row of cells; once more he saw his room, of when he was a boy, and saw himself in bed—next to his bed the lute, and it began to play: he knew its sound and song. The brain was wider than the sky.

But then he thought: Inside that boy, too, was a brain, a brain that would contain another universe, the green expanding universe of his youth, and within that universe were other boys, and other brains, each with its lute and with its rising sun. The sky contained a billion brains, he thought, every living brain was like a burning star.

And yet each brain was just a trifle in the vast furniture of the world—a quivering jelly fitting inside a cup of bone, a tiny loaf covered by a hat, a poor sponge soaked with just a glass of wine—a fist enough to break it. How could a brain contain the sky?

The brain painted the world with color, he once had thought, made it alive with sound, and gave it taste and smell, but now he felt the brain did more than paint: the brain dreamed up and forged all things that were—lutes, rooms, mountains planets and stars.

My brain gives birth to what is real, he thought, to the bloom of the bulrush, the flakes of the pinecone, to the berry of the juniper. It gives birth to the drone bee, the sea grass, to every object large and small, to the meadow nearby, and to the distant peaks. It burns, it glitters. And it has no name. It has no name but I.

He thought: What is, is what can be perceived. Reality is only made of pure experience. The brain can hold the sky because it can beget the soul, and when a soul is born, a universe is delivered.

But then he knew nothing was gained—for how could the brain generate the soul? A woman can beget a child, and that was wondrous enough. But the child's brain was the true father of the soul, the one who could engender consciousness, giving birth to it every time it woke, or every time it dreamed. Flesh could give rise to flesh, an earthly embrace could grow new seeds, it was a marvel, but it was not a miracle. How could mere matter generate mind? It was a mystery, stranger than an immaculate conception, an impossibility that defied belief. Perhaps there was a special part of the brain, an inner sanctum where consciousness's conception was celebrated. Perhaps there was a pivot point where transubstantiation could transpire, not from bread to body, but from brain to soul.

Perhaps, but Galileo's thoughts lost wind and stalled in midflight: for his was an old brain now, gray like his hair. What would befall his world when his brain was put to rest? When the light was off inside his head, would darkness take along his friends, his house and country? Would his memories be lost forever—would everyone and everything vanish—would all be lost? If all was born and buried somewhere in the brain, then when the brain would die, the universe would vanish too.

Galileo's spirit had sunk, but then he heard a distant choir, and in the dream it seemed to speak to him:

"Dull is the brain, its center thin, and shadows only come together there; but the soul, the soul wants more than just a place to dwell—the soul is not a point at all."

So Galileo took breeze again. Perhaps the choir was right, perhaps the soul is just a guest in the dark palace of the brain, a guest who only spends a few nights there, a wanderer who does not wish to settle, but like a gypsy roams without constraint, free soul with a free will, not subject to the mechanism of matter.

And then he heard a voice. "Only what is unhinged is free," said the voice. "You hope to find freedom's lost keys among the chains and locks that lie entangled in the brain? You hope you can distill an insubstantial soul from the foul sewage that rots inside the skull?"

The voice came from a wooden chair that hung from a gigantic scale. Who was the man in the chair, holding a thermometer and a pendulum in his hands? And then it occurred to Galileo: it was Sanctorius, the doctor who spent his life on the scale, thinking all could be weighed, and all should be; Sanctorius, who had borrowed Galileo's instruments to measure, not the sky and its celestial bodies, but the human body and all its earthly fluids; Sanctorius, who, weighing all that goes into a man and all that leaves him, discovered that something was missing from the balance—the warm vapors of the body, which unaccounted, silently, unknown to us, steam out of the pores of the skin, and leave the body dry. Perhaps Sanctorius had discovered the soul, thought Galileo.

But from his hanging chair Sanctorius laughed. "You wish to know what the soul truly is? Only by measuring can it be known," he said, "and I have measured all my life. So I will tell you what I found." Sanctorius laughed again. "Three men and women I weighed with my scale, just before death and after the soul expired. But no, their weight did not diminish, not by the least amount—nothing ever left the body, not even a ghost of vapor. There is no room for the soul in any place on earth. There is no soul, Galileo—there is only the body—and the body is an old greasy machine."

From nowhere, the choir resounded again:

"Light is the soul, and hard to find, hides in the brain as if it were a Naught; ah, the soul, the soul must be the weight of God—the soul is just no weight at all."

NOTES

There is no report of this confused dream in Galileo's papers—most likely it was made up by the author. That would explain the quote from James Merrill's "b o d y" a couple of references to Emily Dickinson's much-abused "The Brain," and one to Gabriele D'Annunzio's "Meriggio." Regardless, like most dreams this one is a mixture of past and present, of the familiar and the uncanny; the scene shifts suddenly, thought follows loose associations, and people and voices (Sanctorius and the choir) appear seemingly out of nowhere. The feeling of flying or floating is common in dreams, though Galileo's account at times seems like a near-death experience: the out-of-body aspects and the sense of entering a region of utter darkness are telling signs. Perhaps Galileo had a brief cardiac arrest: both flying dreams and out-of-body experiences are due to insufficient blood to a region of the brain underneath the temple.

The main preoccupation underlying Galileo's dream is clear enough: Does consciousness (which is to say the soul, since dreams make no fine distinctions) spring out of the matter of the brain, and how? Just as clearly, Galileo has no clue. In a famous passage of *The Assayer,* he had written: "Concerning sensation and the things that pertain to it I claim to understand but little, . . . therefore I leave it unsaid." Having recognized his ignorance of subjective properties, Galileo wisely chose to study the objective properties of bodies. That is why Galileo is often credited with having eliminated subjectivity from the study of nature, replacing it with mathematics and measurement. In the dream, however, he seems of two minds: Is consciousness generated by the brain (and how on earth can the brain do it), or does it exist as a disembodied soul? In *The Assayer,* Galileo had drawn a famous distinction between objective and subjective properties of bodies: "If ears, tongues, and noses were removed, I am of the opinion that shape, quantity, and motion would remain, but there would be an end of smells, tastes, and sounds." But in the dream Galileo, like Berkeley and Kant after him, begins to wonder whether the so-called objective qualities—the entire universe, indeed—aren't also a product of consciousness. Not surprisingly, he seems to be of two minds on this issue, too.

As for the pictures, Comet Hale-Bopp was seen from the Space Shuttle *Columbia.* The 1999 total solar eclipse in France was photographed by Luc Viatour (www.Lucnix.be). The view of the abbey at Vallombrosa, where Galileo studied and played the lute (*The Lute Player* by Caravaggio is at the Hermitage, St. Petersburg), is by Louis Gauffier (Musée Fabre, Montpellier). *The Soul Hovering over the Body* is by William Blake (Tate Collection, London). The chest and nostrils are of Mantegna's *Dead Christ.* The *vanitas* still life (with a subtle change) is by Pieter Boel (Musée des Beaux-Arts, Lille). The cupola of the Sagrestia Vecchia in Florence is attributed to Pesello. The portrait of Galileo as an old man is by Justus Sustermans (Palazzo Pitti, Florence). Sanctorius Sanctorius taught theoretical medicine at Padua, where he met Galileo. He adapted the pendulum to measure pulse rate and invented the clinical thermometer. Living on his scale for many years, he did discover *perspiratio insensibilis*—the evaporation of water from the skin. The illustration of Sanctorius in his steelyard chair, about to weigh himself, is from his *De statica medicina* (1614). In the early 1900s an American doctor, Duncan MacDougall, weighed six dying patients on a scale not unlike Sanctorius's, just before and after the moment of death. He claimed to have proven scientifically that the soul exists and is measurable. He repeated his observations in dying dogs and found that their weight, by contrast, did not change.

PART I

———

{EVIDENCE}

Experiments of Nature

2

INTRODUCTION

Displacements

Knowledge and wisdom begin at the sickbed, Sanctorius had said. A small crack in the body's frame will shake the soul's backbone, a trumpet blaring what we stand to lose.

A man was sitting at Galileo's bedside, tall and dressed in a maroon jacket—his long eyebrows stood out straight as he spoke. Then he introduced himself—his name was Frick—and began telling Galileo what was to come. He would read books of many kinds, encounter people of many ages, and there was much for him to study—about the world, about man, and about the soul. Along the way, he would learn a story filled with revelations—more plentiful and astonishing than in the Bible. Unlike the Bible, this story did not fancy a special place for us—not for our world, not for our race, and not for our soul. It was a story that promised no retributions, good or bad, a story for grown men, not a consoling tale for children, or so Frick said.

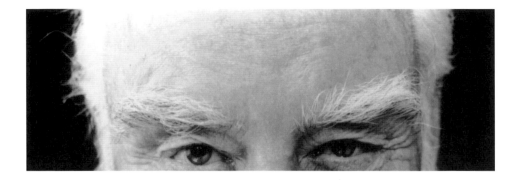

The first great revelation was one for which Galileo had been the prophet—it was about the earth's place in the universe.

Galileo and his telescope had been right—the earth is a small planet orbiting around a common star. Powerful lenses had revealed more, and worse: that our star is one out of billions within our galaxy; that our galaxy itself is a modest one, one out of billions of other galaxies; and that all of them were born to be cast off in a giant universe, perhaps one out of many, one that had been growing for billions of years, and after billions more would die. Even the prophet should feel dizzy at the immense magnitude of the earth's insignificance, Frick had said.

"There are innumerable suns, and an infinite number of earths revolve around those suns," Bruno had claimed before burning at the stake. Perhaps he had been prescient in his madness, thought Galileo.

Though man had risen high and landed on other planets, he had no hope ever to visit the center of his universe, if there was one, Frick had said. The prophet was but a forlorn curate in a remote village of a great empire, one who would never, not in a million lives, travel as far as Rome—if indeed there was a Rome—or even glimpse at the great capital from remote distances. And news, if it ever made it to his obscure corner, would be thousands of years old when it would reach him.

Science had grown and prospered, though whether it was all for the best remained to be seen, Frick also had said. Man had mastered many of the laws that govern nature; some Galileo had surmised, others he could not begin to imagine. Science had given us great power, power to move fast, to generate vast amounts of heat, to create new crystals and new metals, and to send words and images over the earth in little time. But there was no escaping what these discoveries told us: that we are confined forever in a faraway province of the universe; and to that universe our lives are rounded with an instant, drowned inside a point.

Galileo shivered, as he thought of himself watching the night sky on his terrace—young and eager, and did not know whether he felt proud or humbled.

The second revelation dealt with the place of the human race.

That place, too, was in no way special, Frick had said. For we had learned that man descends, through a long lineage of species, from the simplest forms of life—so small and primitive that the first ones came together spontaneously, out of molecules of the earth. And over millions of years, most of the species that had touched the face of the earth had died off, never to return. Others survived for a while by changing and adapting to the harsh ways of the world. Closest to us among the survivors were hairy brothers living among the trees, but even snails and flies were not-too-distant relatives.

Perhaps a poet had said it long ago, thought Galileo: *"The earth created many monsters then, which came out with extraordinary faces and bodies: creatures with no hands or feet, mutes without mouths, blind beings with nothing to look with, and some with arms and legs stuck to their bodies so, that nothing could they do nor go anywhere. But all is to be right for species to last and reproduce—and many have died out. Wherever you see a creature that survived, it was craft or strength or speed that saved it."*

The body of man was not the masterpiece of an all-seeing engineer, but had developed by blind trial and error, over millions of cruel

and wasteful years. The instructions to build that body, revised many times by chance and fate, were stored in simple molecules, threaded together in long, twisted strands. Atoms make up molecules, molecules make up cells, and cells make up the skin, the muscles, the heart, and yes, the brain. Nothing but atoms and the void.

As the poet had said, thought Galileo again: *"All nature, then, as self-sustained, consists of twain of things: of bodies and of void in which they're set, and where they're moved around."* Billions of atoms arranged into molecules arranged into cells make up our whole body, and that of every other man, and that of every other animal or plant. There is no magic anywhere: all is mechanical.

This knowledge too has given man great power, power against disease and death, Frick had said. But there was nothing special about man, his body, or his brain: if one wanted, scores of newborn Galileos, each young, each strong, each feeling a private surge of ambition, a goal set in his fierce mind, ready to fight each other in battles of equal intellects, could be generated out of a shred of his old man's skin—the instructions that mattered were all there. You are merely another beast in the great zoo of the universe, Frick had said.

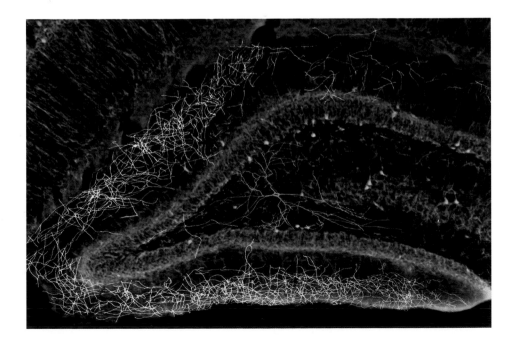

And finally there was the third revelation—the most daunting of all—the one that dealt with the place of the human soul.

The revelation was this: that you, your joys and your sorrows, your memories and your ambitions, your sense of identity and free will, are no more than the behavior of billions of nerve cells in your brain, and of the molecules that make them up. You are nothing but a pack of neurons, nothing but a pack of molecules, Frick had said.

This Frick had said. Perhaps with a tinge of satisfaction, thought Galileo.

And Frick had twisted the knife into the wound. You are proud of your strong intellect, so strong it challenged the stars. But all there is to it are well-endowed nerve cells somewhere in your brain, just like all there is to a brute are bulkier cells in his muscles.

You pride yourself that you are a pious, compassionate man. Instead it is just this: by chance, within your brain, some cells are less loosely knit than those of wanton criminals.

You are sure your will is powerful, your choices are guided by your conscience. Instead you are a mere servant of hordes of nerve cells—you follow their instructions to the letter, Frick had said. You are empty, Galileo, and have no spirit: nothing enters your immature body at conception, and nothing leaves your carcass at death. You are but a slave chained to a dying machine. Over it you have no power, and its end will be your end. It will be the end of an illusion.

Frick's long eyebrows were peering at him, as if to check whether he had made an impression.

That the earth stood still at the center of the universe was indeed an illusion, thought Galileo. Perhaps it was an illusion, too, that plants and animals and the human race had been created and were immutable and perfect. After all, even the poet long ago had said that they evolved by chance and survival.

But the soul? How can the soul be an illusion? How can consciousness be just a mechanical play of atoms and molecules? Because I may be mistaken about *what* I see or feel or think or wish, but *that* I see, feel, think, or wish I cannot be mistaken. I thought I was in the monastery of my youth, and I was mistaken, as I was having a dream. I may be having a dream now and be mistaken again. The world, myself, all life, all history, and all science may well be images and thoughts happening in a dream. But the dream itself is real. My consciousness, whether I am dreaming or awake, is real. If consciousness is an illusion, then only illusion is real, and the rest is conjecture. There is no explaining consciousness by atoms and the void.

Or is there? Sanctorius and Frick, both of them, whether imaginary or real, in dream or in reality, had told him the same thing. Somehow our soul, our consciousness, our world, all is generated by what's inside our skull. How could this be? If there were ways to know, Galileo should rise and inquire. And the man with the long eyebrows stood up to lead the way.

NOTES

Galileo may be excused if he tends to reinterpret facts, events, and even poems in the light of his limited knowledge. That we are not at the center of the universe, we have become used to, perhaps consoling ourselves that while our location may be peripheral, our mission is not. That we may descend from simpler forms of life, we have also had to accept: perhaps we can be proud of how high we have risen from such a lowly start. But that our very soul may be not just mortal but mechanical, the mere whistle of the steely locomotive, as Thomas Huxley, Darwin's friend, once said, may be more difficult to swallow. Huxley, who was sure that evolution by natural selection could explain most things human, was at a loss with consciousness and left the door open: "How is it that anything so remarkable as a state of consciousness comes about as a result of irritating nervous tissue, is just as unaccountable as the appearance of the Djin, when Aladdin rubbed his lamp."

Galileo's guide in the first part of the book is Francis Crick (his is the first half-portrait). After discovering DNA's double helix, Crick devoted his life, together with Christof Koch, to the study of the brain basis of consciousness. Unlike Huxley, Crick left no open door for the soul, not even a window of opportunity: "You, your joys and your sorrows . . . are nothing but a pack of neurons" is from his *The Astonishing Hypothesis: The Scientific Search for the Soul* (Scribner, 1994). Koch, on the other hand, may be less categorical.

As for other pictures and quotes in the chapter: Dürer's wood-cut *The Revelation of St. John* (*9. St. John Devours the Book*) is at the Bibliothèque Nationale, Paris. Giordano Bruno, an early upholder of Copernican views, claimed that the universe was infinite in *De l'infinito universo et mondi* (London, 1584). The Cigar galaxy, M82, with a supergalactic wind, is from "Astronomy Picture of the Day," NASA, April 25, 2006. One wonders whether the eyes of the scallop are more or less enigmatic than those of the *Mona Lisa*. (Scallop eyes are from *Animal Eyes* by Land and Nilsson, Oxford University Press, 2002.) The picture of the hippocampal formation is courtesy of Gyuri Buzsaki. Each yellow triangular-shaped cell body is a single pyrami-dal neuron. Like a tree, each neuron sends down a long root (axon) by which it transmits signals to other neurons and sends up a thick set of branches (dendrites), where it receives connections (synapses) from the axons of other neurons. The human brain has about one hundred billion neurons and at least a thousand times more synapses among them: its many stars are not floating in the void. The poet quoted by Galileo is Lucretius, who in his *De rerum natura* (*On the Nature of Things*) attempted an exposition of science in poetry. This is a genre that has not had much success since.

3

CEREBRUM

*In which is shown that the corticothalamic system
generates consciousness*

Through me the way to the city of woe, through me the way to eternal pain, through me the way among the people lost, thought Galileo when he had crossed the gate.

The narrow hallway, dimly lit and empty, that led away to an endless line of doors, seemed to belong to some forgotten hospital. Galileo went up a narrow ramp of stairs, still following the man named Frick, and saw another corridor. The first door let light seep through the sill, and Frick opened it softly. There on a narrow bed lay an old man, his posture strangely crooked. Nearby was a book that showed this title: *De revolutionibus orbium coelestium, 1543,* the year Copernicus had died.

Galileo remembered what Copernicus had written in the preface of his book: in his view, the universe was an orderly whole, in which displacing any part would disrupt the entire edifice. Instead, the views of his predecessors were like a human figure with arms, legs, and head put together in the form of a disorderly monster.

But now Copernicus himself was like a monster, arms and legs bent and stiffened like the gnarled branches of an old oak tree.

At Copernicus's feet a woman knelt, her face turned to the ground. "She is always there with him," said Frick to Galileo. "She sees him breathe and hopes he is going to resurrect." He looked at her: "Woman, nobody can resurrect if the brain is dead. There is nobody there, woman, nobody is home."

The woman raised her head: "How can you be so sure, stranger? His face once smiled at me—it was night and I held up his head to drink. Sometimes he moved his lips, as if to speak. And then he smiled—or so I think—for I had never seen him smile before."

Copernicus lay without moving, but Galileo saw he breathed quietly. "Blood has burst into his brain and destroyed the cerebrum," said Frick, leaning with outstretched arms on Copernicus's bed. "But he can still breathe, and his heart still beats, because the lower brain was spared. As if breathing were living," Frick added almost imperceptibly.

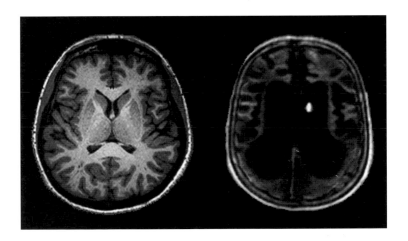

"The vast cerebral cortex and the precious thalamus—the small bed of neurons where the cortex crouches—the entire corticothalamic system has been destroyed," explained Frick. "The cortex, you see," he said, peering at Galileo, "is a sheet as large and thin as this bedcloth—and just as convoluted. And the cloth is truly an immense forest, which covers every ridge and valley of the brain." As in his dream, thought Galileo. "Each tree is a nerve cell," Frick went on, "and just as trees are densely packed in groves, so neurons often band together into groups, each containing maybe one hundred of them. These small groups of neurons, you see, are the building blocks of the brain and send signals to each other at great distance through thin wires." Or so he explained.

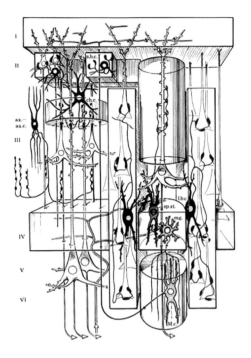

Then Frick went on: "You see, each small group of neurons within the cortex, each one of them, has its own special function to fulfill: groups of neurons at the back of the cerebrum deal with sight, others along the middle with sound, yet others with touch, smell, and taste; and groups of neurons at the front of the brain deal with thought, or with emotions like anger and joy. But specialization goes even further," Frick said. "Among the groups of neurons at the back of the cerebrum, some care for the color of objects—can tell with perfect ease if something is red or yellow, but could not care less if that's a beet or if it's a lemon, in fact they have no idea. Others instead are particular about the shapes they like—one may like the pyramid and one the sphere, but to them red or yellow are all the same. Others still care about the way things move, indifferent to their shape or color. And among the latter, you see, some care just for sideways movements, others just for movements up and down," said Frick, moving his finger in front of Galileo's eyes.

Galileo thought of the translucent mass, gutlike in its folds, that he once had seen on the anatomical table in Padua. And then he thought of a great and lively city, where a different guild or craftsman could be found for every possible need, lens makers and ear trumpet makers, makers of all fashion of clothes and makers of perfumes, and winemakers and cooks, and geometers and mathematicians and logicians, and great orators and poets and artists and musicians.

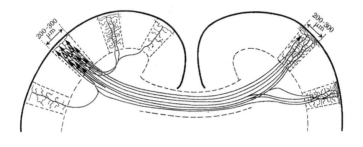

"But just like in a city the members of different guilds and professions must talk to each other, and exchange orders and goods, so do the elements of the brain," said Frick. "Much of the cerebral mass is made up of thin wires through which its specialized elements talk to each other: always pushing and pulling, forming coalitions that do not last for long, then changing alliances all the time, like the many factions inside Florence. The wires are so many," he said, "that their length is greater than all the roads of Italy. Where they come together in great numbers, they form the white matter of the brain, a thick fat mesh that lies underneath the bedcloth of gray matter. Without those wires, the brain could never function, just like a great city would come to a halt if its roads were blocked."

Then what has happened to the brain of Copernicus? asked Galileo. Has the flooding blood destroyed the elements that formed the governance of the brain, the prince who oversees the work of the great city, and all its counselors? Or has it lost some very special element, the one that was responsible for consciousness?

But Frick said in a loud voice: "The brain is a democracy—there is no such thing, in the brain, as a prince or pope, who sees and hears everything, and takes all decisions—no privileged seat of consciousness, no pontifical seat. Consciousness needs the cooperation of many specialists, each one providing its unique contribution. So if an illness destroys the regions of the brain that are specialized in telling colors," said Frick, "you will become color-blind—the sun will turn white and the sky gray; if those specialized in recognizing faces, you will not know your children when you see them; if yet other specialized regions are ruined, you may not be able to perceive movement, or to sense emotion, or to understand language, or speak, think logically, or distinguish between right and wrong. But you lose each time only one part of consciousness, not the whole," said Frick.

"To lose consciousness altogether, the damage must be large," he added, "either because most of the regions of the cortex are dead, as

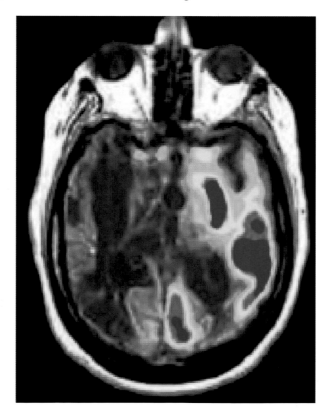

they are with Copernicus, or because the wires through which they talk are faulty or broken. Sometimes even small injuries can cause great havoc, especially near the middle, deep inside the cerebrum. That region is like a hub, governing the traffic among all the others, and if you interrupt the traffic, then naturally the edifice of consciousness collapses," Frick said.

"But sometimes, when blood gushes through the cerebrum, like a river in flood taking all life with it, a few elements may survive, a small, shipwrecked island in a sea of waste. Consciousness vanishes, but some individual function may remain, as if a poor cobbler alone were left to whine in a dead city."

Copernicus lay mute: they called his name, asked if he could hear, if he knew where he was, if he felt any pain. No question would make him say a word. Arms and legs jolted back when Frick touched them with a pinprick, but when Galileo made a threatening gesture, Copernicus did not respond.

And yet his eyes were open, thought Galileo, and roamed within the orbits. Did any life revolve with them, did anything move inside? Did sparks of thought still stir?

"At times he yawns, as if he were tired, or he may grunt, as if he wanted to stretch, but those are mere reflexes, one nerve pulling another in the lower brain," said Frick. "What we have here is an awake coma—a vegetative state."

Inert, an old, contorted plant, thought Galileo.

"Some people do remain like that for years," said Frick. "Indeed, no more than plants—like trees their bark does age, but they don't grow any wiser, as they experience nothing, and there is nothing for them to sense and learn."

Transformed into a plant, as in a cruel Greek myth, thought Galileo.

"A plant!" exclaimed the woman, who had kept silent until then. "Which plant would bleed warm blood, as he when he was felled? Which bark does hide a beating heart, as his that speaks to me across the chest? No tree would turn to watch the hand that combs his hair, as he once did. No tears did ever flow to streak a flower's skin, unless it was the dew, but dew was never salty of sorrow."

"Woman, you just project your wishes onto an empty shell," said Frick with a stern glance. "What you should do is take care of yourself—do not grow old and tired waiting upon a corpse."

As if talking to herself, the woman muttered: "The bishop banned me from this house, but why should love by force be wed to guilt, and end not in communion but divorce? The bishop warned me of misfortune, though I thought he meant the blow for me alone—oh God, is life suspended reward for life in sin?" She paused and whispered: "Ours was the kindest sin of all."

"Woman," said Frick, "your only sin, you see, is the sin of projection. You wish those eyes would lead to a soul that needs your help and hears your voice. So much you wish it that now you fill his empty gaze with sight, and in a grunt you hear the echo of his words. Listen to me: behind that stare there is no other soul but yours, reflected in a mirror, a mirror that on the other side is blank."

The woman remained quiet, so Frick went on: "You know the sin I am speaking of—the same as when one falls in love: all men desire beauty, virtue, and grace. Just give a man at the right time, when certain humors are receptive, a specimen of woman—a sketch that's rough and raw—he'll dress it head to toe with all his wishes, virtues that don't exist but in the imagination."

Then Frick smiled at Galileo. "Some fish project their yearnings onto an oval shape of wood, as soon as it is painted red, and mate with it for life. At seeing such fish we laugh, but then who are we to laugh? The walls of our aquarium we can't even see, and our own humors hold us by the chain. Believe me, the sin of projection is the original sin. Men saw the lightning, heard the thunder, and felt the earthquake: so what their heart would fear, their mind could not explain, they placed in Heaven and they called it God."

The woman turned her head to Frick and looked him straight in the eyes. "You think you may know all, stranger," she said. "What if your science, too, is just the image of a need—the need that things are clear and solid and all can be explained? What if your science, too, is just another kind of church, projecting down the earth and up the sky?"

Then she stood up. "Look at me," she said. "Anna is my name—Copernicus's own Anna. He saw in me a beauty I never had (or so the others thought). Yes, stranger, you are right: it was indeed his love that forged me as I am. So in return let me imagine his soul, and see it shine behind his eyes, and read a smile between those silent lips. Let me plant words for him to grow for me. You are right, stranger, but not the way you think: what is, is what can make things happen. So now his soul, the soul that's resting in that body, is filling me with love and grief at once, and thus it must exist."

How could Frick reply to that? "Reality and appearance are at war," he said eventually. "They always are. This woman, you can see, has made her choice," said Frick, turning to Galileo.

True, reality and appearance are at war, repeated Galileo. Though it is not clear which side you'd want to be.

Then Anna took Galileo's hand and pressed it hard. Like a broken machine, Copernicus's mouth had been moving, on and off, though it whispered no sound. But if they watched more carefully, perhaps they could make out his words, their shape without their sound. Their shape, thought Galileo, was saying: *"E pur si muove."*

NOTES

The initial reference, quite appropriately, is to Dante's *Divine Comedy, Inferno,* Canto III, which Galileo knew by heart. It is not clear whether the gate that is mentioned here is to a hospital (as Galileo surmises, though one without doctors or nurses), or rather a monastery, or a prison (the cloister is of Fontenay Abbey, Montbard, France). What is certain is that Copernicus died before Galileo was born, so unless his vegetative state lasted much longer than was possible at the time, he and Galileo could not have met. Nevertheless, Copernicus did die of cerebral hemorrhage, losing consciousness and slipping into a vegetative state until, shortly before his death, he may have briefly seen his book finally printed. Copernicus's portrait is from the Copernicus Museum in Frombork, Poland. Anna Schilling was Copernicus's housekeeper, and Bishop Dantiscus ordered her removed from the household of the astronomer/priest. When she says, "What is, is what can make things happen," she is apparently enunciating a principle of causal ontology, which sounds a bit surprising coming from her. On the other hand, all the talk about "projection," a dubious psychological defense mechanism that in Frick's words assumes almost universal significance, seems just as preposterous. The painted Mary Magdalene is from Masaccio's *Crucifixion,* the carved one by Donatello. The statue is Bernini's *Daphne* (with some changes), at the Galleria Borghese in Rome. The chaste nymph Daphne was turned into a laurel tree, pursued in vain by Apollo, god of light. Ovid says that during her metamorphosis, when bark was already covering most of her body, Apollo's hand could still feel her heart beating beneath it, just as the hearts of vegetative patients are still beating despite the fading of consciousness. The drawing of a small group of neurons arranged in a column, also called a minicolumn, and of two minicolumns in the right and left hemispheres talking to each other through neuronal fibers, is from the anatomist János Szentágothai (Ferrier Lecture, 1977). The pair of brain scans are from a healthy subject (left) and from a patient in a vegetative state (right), in whom most of the brain has been destroyed (black areas). The last scan is from a paper by Niko Schiff and others: "Residual Cerebral Activity and Behavioural Fragments in the Persistently Vegetative Brain," *Brain* (2002). There are indeed patients who are left with just an island of partially functioning brain, which can at times produce an isolated word, or an isolated movement. *E pur si muove* (And yet it moves), so the story goes, is what Galileo mumbled when leaving the Inquisition that had found him guilty of heresy, forced him to abjure his Copernican views, and put him under arrest for life.

4

CEREBELLUM

In which is shown that the cerebellum, while having more neurons than the cerebrum, does not generate consciousness

"Most of what happens simmers beneath the surface—unseeable by the mind—most of the action unrolls in murky waters," said Frick, and threw the door wide open.

Inside the room, in front of a large canvas, a man was standing with legs wide apart, and his hand shook when he moved the brush.

I know the painter, said Galileo to Frick. It is the great Poussin. I know the painting, too. See, Janus indicates the instant between past and future, the instant of conscious present. On the right, Time marks the beats of music on the zither, the rhythm of dance and life. And in the middle, the dancing men and women represent the seasons and the cycle of time.

"Time is not circular, my dear Galileo," replied a figure emerging from behind the canvas. "Perhaps that of the planets is, but not the time of men." It was the Pope's doctor, Rome's Protomedicus. "Look at the putti. The right one holds the hourglass: inexorable is the passage of time, it says; the left one says that consciousness is short-lasting, like a bubble of soap."

Galileo did not reply. Instead, he whispered something into the doctor's ear. "Trembling?" exclaimed the doctor. "Of course I know, Poussin has been this way for some time now—his brushstroke clearly has suffered. But rest assured, his health is fine. I've seen another case like him: a young woman with the same kind of trembling, who kept her legs apart when standing. When she died of childbirth, I examined her brain. In the back of her skull, under the nape of the neck, something was missing: there was no cerebellum—that furrowed brain that looks like a small cabbage, or like the tree of life. Our Poussin, too, must have lost his cerebellum, perhaps it was a stroke. It seems that all it does—this beautiful little brain—is to make sure our gait stays elegant, and our hand shakes not."

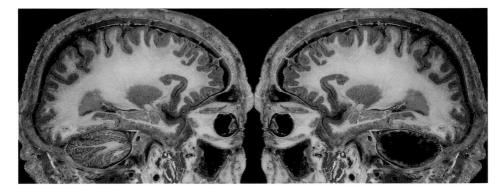

"Quite right," said Frick, stepping forward, "though it does other refined things. It keeps track of our movements much faster then we can, and it corrects and stabilizes them. It is a piece of brain as intricate as the cerebrum—in fact, it has an even larger number of nerve cells. And like the cerebrum, it receives signals from the senses and controls our actions with perfect accuracy."

Galileo was thinking. The canvas was grand, the subject noble and indeed replete with meaning. But it was singularly immobile.

Although it was about Time, Time seemed frozen: everywhere pointers to time and movement, even a dance, but real time, the time of the pulse, the time of the pendulum, that time was missing. It was as if the painter's pulse and pendulum were stuck.

Right then the painter turned toward them. "We must not judge by our senses alone, but by r-reason," he said with a strange, wavering voice. "My hand m-maytremble, but my mind does not; for things preserve their being irrespective of accidents. Go-gobeyondthesurface, and contemplate the idea: the cone, the cylinder, and the sphere." He paused, took a breath, and said: "The p-purest are the idea of good, and that of truth. It's just that . . . ," and there he paused again. "It's just that those are far too hard to paint."

"Well said," said Frick. "Many of the neural systems in the cerebral cortex do just this—they learn to predict what remains constant in the world, despite the seeming onslaught of constant change. They paint a scene of what the world should be, much as you paint it, with scarce regard for all the changing details our senses bring in most of the time. So in our consciousness the cone's shape stays the same, though when we see it from different angles, the images formed onto our eyes are different entirely. So the color of ripe fruit stays the same under the warm light of sunset and the cold one of lightning, though the light reflected by its surface doesn't. Yes," said Frick, "the scene in which we live is an abstraction, experience must be make-believe, a painting by some clever master. Reality may swirl in whirlwinds of irrelevant, superficial change. But information is in what is constant, and general, and deep. That kind of information is what consciousness requires. Isn't that also what art is trying to capture?" asked Frick.

"I paint what I know, not what I see," said Poussin.

"But what if what you see is what you know?" asked Frick.

"Why don't we leave alone all these abstractions and examine the trembling that bothers our Poussin?" said the doctor.

"Quite right," said Frick. "Poussin's cerebral cortex, the one that paints invariants and abstractions, imagines and predicts his view of the world, the one that generates his consciousness, surely is still all right. It is his cerebellum that has failed, a system that cares not about the constant, the general, and the deep, but needs to know exactly how things are, here and now, in exhausting, accurate detail—to calculate how far fingers have to open to grasp the brush, reckon how far to move the eyes to reach the canvas's edge, how much the arm must stiffen to steady his stretched hand, how much the trunk must tighten when the shoulder lifts to draw. Such systems do not need to paint a grand scene of the world, they do not worry about what it all might mean. Their many parts don't need to talk to each other, they only need to do their special private job, to do it right and fast, oblivious to the rest, uncaring of the essence. You cannot have a prefect grip and grasp the universe all the same."

"You cannot grasp the universe, indeed," intervened the doctor, turning toward Poussin. "Then let's make sure at least that everything else is working: My dear Poussin, anything wrong besides the tremor and the gait? How are your sight, your hearing, and your touch? Your smell and taste? Has your grip stayed strong?"

The painter scrutinized the physician. "Though you may be a great doctor, your eye is not as good as mine. For all is fine with me: I see colors and shapes, I hear the sounds, I touch and taste as I always did. In the g-greattheater of my mind, the p-play is still going on. True, doctor, there are things I cannot do: I cannot play, and I cannot dance, and so I paint them. But if the gait of my thought may be lame, doctor, it still walks faster and further than yours." He pointed to a finished canvas: "Tell me this, G-Galileo: Is there more truth in his dissections or in my art?"

But Galileo was not paying attention: his mind was weighing the cerebrum against the cerebellum. When the cerebrum is destroyed, a universe is destroyed, but when the cerebellum dies, nobody is dead: Copernicus was gone, Poussin was painting. So the cerebrum is necessary for consciousness, the cerebellum not. Yet the cerebellum is as much brain as the cerebrum, as rich and well endowed, and trades as much with the world outside. You told me the cerebrum was like a great city, he said, looking at Frick, but so is the cerebellum.

"Indeed," answered Frick, "now that I think of it, I am not sure myself why consciousness should be so fussy. The cerebellum does not lack nerve cells, or any other ingredient that makes the brain a brain. Perhaps it is the building plan," he said after a brief reflection. "Cells in the cerebrum are all connected, directly or through a few intermediaries, as I have told you. They talk to each other all the time. Cells in the cerebellum may lack the right connections—what they receive, they process and send out, but cannot talk to each other."

"My studies may be outdated," said the doctor, "but what you say reminds me of an old tale, the tale of the two cities. Listen," he said.

Once there was a king who reigned over a great city. He was rich and powerful but was afraid of what people would say to each other. He knew the citizens did abide the law and did respect their king. But, thought the king, how could he know what they said to each other when nobody saw them? So he hired some trusted men and had them keep an eye on the citizens and write down carefully what they heard.

The men went to work and began filling books and books with all the things, great and small, the citizens would say. But that did not work well. Nobody had time to read the reports, though a large department had been created just for that, and even so, how could one know whether the citizen said one thing and meant another? Besides, who knew what happened when the informers went to sleep? So the king had them do the shifts, and when one went to work with bleary eyes, the other went to bed with aching arms for too much writing. And he made sure the informers themselves would not be left unchecked—so he set up, in great secret, a second tier of informers, who informed on the first tier. That will do for the moment, thought the king.

But soon the king realized that he had been careless. The informers would inform when people gathered in the streets, but who could know what passed between husband and wife? And not just at the dinner table, but when they huddled in the hideout of their beds? Or when a mother told her daughter the secrets of marriage, what other secret might she confide? There did not seem to be a solution. The king convened his counselors, and none of them had any good idea. But then the court jester raised his hand—his name was Modulus—and made a suggestion the king thought was foolproof.

So the next day the king's masons and carpenters went to work, cheerful because their wages had doubled, and in little time carried out the king's orders. When the oldest mason had finished his last assignment, the king surveyed the city from the highest window of his highest tower and was happy with what he saw. For everybody, from the oldest mason to all the other masons and all the carpenters, from every craftsman to every landlady or maid, from every child to every granny, everybody was safely confined within his own cubicle.

Each cubicle had thick walls, strong because there were no windows. Inside the box was everything one needed—a bed and a light and running water, and food was provided every day through a little door, so small that even children could not slide across. The jester had truly thought of all—the king's dogs would carry everything to and from the castle. So through the little door back went the work done by the craftsmen—the miniature sculptures for which the city was famous, the knives forged by the blacksmiths, the jewels fashioned by the goldsmiths, the clothes embroidered by the needle workers, even the food the maids had cooked.

All was good and safe, thought the king, and indeed he heard no complaints. He could hear instead the industrious hammering of the craftsmen in their boxes, and the clangs of pots and pans made by an eager cook in her cubicle; otherwise the city was veiled with pleasant silence. Just as the jester had said, now he could stop worrying about what people would say, as there was no way they could talk to each other. And so he prepared for a long, serene reign, and from high above, he watched the endless plain, dotted with innumerable cubicles, and inside each cubicle one of his subjects was hard at work.

But then one day the jester raised his hand again, saying he had had a dream that an army had entered the city encountering no resistance. How could the enemy be stopped if the citizens could not speak to each other? What? said the king, annoyed. It is enough that each citizen receives his orders from the castle, without a word being wasted. What do citizens need to talk about? Sire, said the jester, if one who spots the enemy cannot alert the others, call them all to arms, to form a mighty legion where there is great need, we cannot survive, not against an enemy who, I've heard, devours our dogs alive, raw from the bone.

So they went, the king and the jester, down to the city, to the vast plain, and stopped at the very first cubicle. The king knocked at the small door and ordered it opened, but nobody would answer. The king tore down the little door with his saber, and when they entered the cubicle, torn and dirty, they saw the old mason sitting on a large stone. What happened, old mason? they asked. I've lost my mind, he answered. All this time alone in my cell, I must have lost my mind.

Has the entire city gone mad? asked the king. The city? exclaimed the old mason. When I built the boxes, I called our best carpenters and blacksmiths. Oh, they were such master builders! For each stall, sire, for each one of them, they built an ingenious machine, made of wood and iron, which did the simple deeds each citizen must perform for you, a machine that did everything it had to do without exchanging a word with anybody. So for the cook they built a food processing machine, for the cordwainer an automatic cobbler, for the glassmith an automatic blower, and for the butcher a self-propelled guillotine with a carver attached. And for the priest . . .

Then what is everybody doing? asked the king without letting the mason finish. Are they idling in their boxes while the machines do all the work? Because now we need all their help against the enemy, the enemy that is devouring our dogs alive.

But you won't find them, sire, none of them, said the old mason. They all left, long ago they left, and built themselves a new town, yes, where all the time they do nothing but talk, they shout and scheme, trade and argue, and do all sort of things together, stage merry plays and sing loud choirs, and there . . . there, said the old mason, who was beginning to make all kind of grimaces, there they must still . . . , and he stood up tall, and it seemed that his cheeks were overblown and close to bursting . . . there, and he could not suppress it further, there, he burst out, they must all be laughing, laughing about you—like happy drunkards they are laughing, and their laughter fills the town hall day and night—and no sooner has a joke reached the end of the hall than another joke starts on the side and causes an even louder roar, and they sing together, and the songs keep changing, but they are all about you, lewd songs, and unspeakable things are sung, and those who scream the loudest are the schoolgirls, and old men like me chuckle so hard that no breath is left in their chest. What would I know, sire? I am just an old, crazy mason. And he sat down again. But one thing, sire, is sure, he said in a somber tone: empty are the cells, and the soul has vanished from the old city.

So, thought Galileo, there are two great cities in the brain. The cerebrum, where citizens of all kinds and manners can argue with each other, and as they speak decide on things together. And the cerebellum, a city even more populous, but there instead everyone lives alone, without talking to anyone, in his own cell taking care of his business. That, thought Galileo, may be why consciousness lives in the cerebrum, the great, bustling metropolis, the lively democracy—and the cerebellum is an immense, silent prison.

NOTES

The painting is Poussin's *Dance to the Music of Time,* which is at the Wallace Collection in London. Poussin did develop a tremor late in his years, although an analysis of his brushstroke suggests that it may have resulted from Parkinson's disorder rather than from a cerebellar problem. See "The Movement Disorder of Nicolas Poussin (1594–1665)" by Patrick Haggard and Sam Rodgers, *Movement Disorders* (2000). The explosive speech of Poussin is characteristic of some cerebellar disorders. Poussin's self-portrait is at the Louvre, and so is his mysterious *Et in Arcadia ego (Arcadian Shepherds),* very much the ideal painting of an idea. Poussin's notion of art has clear affinities with Plato's, but so has the organization of the cerebral cortex, as briefly illustrated by Frick. From a flurry of signals that bombard them, neurons in the highest parts of the cortex have learned to extract the constant, the general, and the deep: the higher one goes, the more invariant to irrelevant change neuronal responses become, the more abstract, and the more behaviorally meaningful. And the same neurons force such categories back onto the world, to predict what it might be like: we see what we can imagine, and perhaps that is why we can see anything at all. Consciousness lives on such constancies, abstractions, and depths, and sometimes art does too, or at least Poussin seems to think so.

By contrast, it seems that tasks that require rapid adjustments, are of local interest, and do not require a vast context of knowledge remain outside our broader consciousness: such tasks can be executed automatically and are carried out by dedicated modules of the brain that can perform their job fast, well, and in relative isolation. Indeed, when there is no need for arguments and discussion, when things are automatic and repetitive, they can move into the assembly line: the great lively city of the cortex gives way to the efficient cubicles of the cerebellum. As Frick notes, though, there may be such dedicated systems even in the cerebral cortex. As studied by Goodale and Milner in *Sight Unseen* (Oxford University Press, 2004), and others, patients with lesions in some parts of the cortex lose the ability, for example, to shape and size the opening of their fingers correctly to grasp an object but have no problem reporting consciously the object's shape and size. On the other hand, patients with lesions in other parts of the cortex lose the ability to perceive shape and size, but their fingers go on "seeing" shape and size and correctly adapt their grip. The aerial photograph is of the Algerian city of Beni Isguen, by George Steinmetz.

5

Two Blind Painters

In which is shown that sensory inputs and pathways
are not necessary for consciousness

Seeing is but a combination of old memories, a private dream led by a tyrannical director, shouting his shifting orders from the outside.

This Galileo thought entering the next room, for in that room too was a vast canvas, showing a cabinet of curiosities. A man stood near the canvas, moving his hand over its surface with unnatural slowness. Was this the circle of sick painters? And who was the lady seated nearby, holding her hand on the heart?

"You see," said the man, "this painting is so wonderfully rich—it took me almost two hours to go over it again. The woman watching a painting within the painting must be the sense of sight personified, and the art chamber as a whole signifies visual perception. It is an allegory of vision, you see. It means that light leaves the eye and shines over the objects that surround us, so we can see them.

"At least this is what I once used to think," said the man, taking his hand off the canvas. "Unless the eye is like the sun, it cannot see the sun. But I was wrong. The eye is just a door through which its rays enter our mind. Vision is in the mind, not in the eye."

Galileo remembered what Kepler used to say: vision occurs through a picture painted on the dark surface of the retina—the eye is like a camera obscura, where the image is reversed; and Kepler used to tease his audience to explain why we do not see the world upside down. (Not that he knew the answer.) But if vision is in the mind, thought Galileo, it does not matter that the image in the retina is reversed.

"Indeed," intervened Frick. "The retina has little to do with conscious vision: at its center, where the optic nerve takes its leave, it is not even sensitive to light, yet we do not see a hole in the middle of each sight; and though the outer parts of the retina are blind to color, we do not see a gray sky envelop the gold of the sun; and finally the retina flickers all the time, because our eyes move imperceptibly, but the images seen by consciousness, those do not flicker at all—they are stable and majestic."

Galileo saw that the man was still going over the canvas with his hand, inspecting it in every detail. You too are a painter? he asked.

"At your service," answered the man. "Compà Zavargna was once my name—you may or may not have heard my story. You see, stranger," said the man, "one of my brothers of the Academy (we met secretly under the auspices of Bacchus) made an elixir for me—he distilled boxwood and brewed the mushrooms of the Val di Blenio (what else he did with it I do not know)—which tasted sweet and did enhance my powers: so when I drank of it, my mind expanded, my art as well—colors became more vivid, shapes more real, faces took on a life of flesh, and limbs walked off the canvas, to greet me and cry about their fate. So I began to imagine my masterpiece—*The Fall of the Rebel Angels*—and in the imagination I could hold it, as real as if it were before my eyes, and contemplate each brushstroke, refine here, enliven there, playing with each corner of my creation. Every morning when I was lying in bed, my mind stirred by the potion, I worked without relent on my great *Rebel Fall* and saw it grow to excel all I had seen before—the colors of Titian and the shapes of Michelangelo, the passion of Mantegna and the genius of Leonardo, all fused within a single canvas: the lips I painted screamed, the eyes turned to implore for mercy—the painting was so alive it drew me inside and almost scorched my soul, until I talked to the Creator I had myself created, discepting and descanting of the highest art."

I am not sure I know your work, said Galileo. Where does *The Fall of the Rebel Angels* hang?

"Stranger," said the man covering his eyes, "I see you do not understand, or act as if you don't. Because the morning the *Fall* was finished in my mind, and I woke up to paint it—so sharply that every detail was etched within my consciousness, that all I had to do was put it down on canvas—the morning when I was to be crowned the greatest, having exceeded all and everyone—that morning, when I opened my eyes, I saw my room was dark, and thence my day grew darker. The poison in the drink had taken its revenge—the painting was still burning in my head, but it was locked inside, an unseen prisoner within my mind. That day I did not paint any further."

Was your masterpiece lost forever and for all, asked Galileo, or can you still behold it in your mind?

"There is no greater sorrow than to remember happy times in misery," said the painter. "It lingers in the imagination, but it is not what it was. All life has now been drained away from it."

Then tell me, if you please, said Galileo, the canvas you were touching—the allegory of vision—can you close your eyes and see in your mind the woman watching the painting in the painting?

"Of course I can, stranger. I have touched her more than once, and my friends filled in by telling me what I had missed."

And did they tell you what color drapes her?

"Certainly—it is a turquoise shawl."

Can you see that color vividly in your mind?

"Of course I can see a bright turquoise shawl in my mind. You know a painter must be good at imagining how things look," he answered. "Otherwise we would have to try out all colors on the canvas, instead of choosing them beforehand."

And when you sleep, and dream of flames, can you see their color? asked Galileo.

"Naturally," said the blind man. "Many of my paintings have come to me in a dream. Don't you understand that form embodies all that may be occasioned in the imagination and can be seen through the eye?" Casting his eyes upward, he added in a piercing voice: "I wanted to rival the draughtsmanship of the prince of painters, equal the inventions of the Great Druid. But when I reached the age our Lord died on the cross, fate crucified me to a blind sepulcher. Now both my eyes are rotten through the poison, as good as lost. So I ended up a man who paints, but only in the imagination—a blink and color fades from all my works; I ended up a man who strives, but never can create—a man who writes a treatise on begetting children, but can have none himself. *Osc' che l'oc' nol ve com gal nol cant.*"

The blind painter fell silent, his head turned down. But the old lady, who had not yet moved, now touched him gently on the forearm and spoke: "I feel your anguish in my heart, my boy, I do. For me too it's difficult to paint, they always ask me but I am too tired—you know, don't you, that I am almost a hundred?"

The man recoiled: "Do not remind me, foolish old lady. Your eyes should have been closed by death, long before mine were covered by darkness." He raised his head toward Galileo: "To think that in her day she was the envy of Europe, her works were prized by all the masters. But now like me her sight is paralyzed, and she lives at the mercy of others, a burden to herself."

"Why do you like to claim that I am blind, my boy?" replied the old lady with a smile. "Maybe it makes you feel better, though I do not understand. I still can take good care of my appearance and know how I should turn my head so that my wrinkles . . . What are you saying, old Sofonisba? What would this young man care about your wrinkles, of all things?"

"You and your wrinkles!" said the blind painter. "If truly you can see, why must the maid lead you everywhere? Why can't you recognize your visitors until they speak?"

"My boy," she answered, "an old lady is uncertain in her steps. She must think twice before she risks an answer, and in my house it's dark. They keep the curtains shut most of the time."

"Lady, always the same excuses," said the blind painter. "Always denying you are blind, and yet you cannot see anything at all; indeed, as far as I can see, you are evidently much blinder than I am. For," he went on, "at least I know what blindness is: blindness has been my mistress for far too many years, but from my mistress I have learned much—from her cruel whim I learned that I still can see—I can see space's flight in front of me, the shapes of men and beasts, things blue and red, pale and bright, as fully as if they were painted by Titian. And I still dream of paintings—I see them in my sleep. If I could not see them, how could I write my treatises? But you? You do not even know what it is like to be blind, because you do not know what it is like to see. And with your art, ah, there you are fully lost—you couldn't even tell what perspective is or means."

"Oh yes I can, my boy," replied the old lady. "Perspective is that habit of the mind by which the people who are close to you are larger and more important than those who are distant relatives, or those who are total strangers, who are indeed very small."

"It's beyond help," said the blind painter, looking at Galileo. "Let me try her again. Lady, can you tell me what is painted inside the frame that's on the table that's on the left of the great canvas I am touching? It's a strange object most of my friends at first don't notice, but just by touching its peculiar texture, I can tell what it is, even without the benefit of my eyes."

"Sure," said the old lady. "Nothing could be easier. Let me see, now, what could it be? Well, in a nutshell, in fact it could be anything. Is it a silvery spider web? I am not sure, but it may contain a lot of other things, perhaps all of them. How is it bounded?"

"You see," exclaimed the blind painter aiming at Galileo, "she has no idea. Lady, can you name any of the objects on the canvas?"

"Well, young man, that would be difficult indeed, as it is kind of empty around here."

"Empty?" said the blind painter. "It's as crowded as a marketplace in Naples. How can you say it's kind of empty?"

"Of course, my boy," said the old lady, "that's not what I meant. Of course it is all crowded—if you only knew how much it cost me to furnish the room with those expensive things."

"Can you believe it, stranger?" said the blind painter. "She is blind like a mole and does not know it. She does not know of what she is talking. She does not even know we are describing a painting and thinks we are wasting words about her room."

"I beg you, young man, don't be rude," said the old lady. "You see, young man, it stands to reason, if this is my room, and I spent so much to furnish it with all sort of precious things, then it stands to reason there should be paintings too."

"Something's terribly wrong with you, lady," said the blind painter. "Sight is the great queen of the senses: her realm the greatest in the mind, her possessions more varied and more extended than those of the other monarchs. But her kingdom is in the mind, not in the eye. And this great queen seems to have left your mind an orphan."

Frick had been listening quietly the whole time. Now he neared Galileo and whispered in his ear. "The blind painter is exactly right," Frick said. "He merely lost his eyes, but can still see within his brain. The old lady, instead, has cortical blindness: she lost the portions of her corticothalamic system that contribute seeing to consciousness—her blindness is a blindness of the soul, not of the eye: her eyes may be all right, but she does not know, anymore, what seeing means."

"Just like you would not know what bees might think of flowers?" asked the old lady absentmindedly.

Just so, dear lady, answered Galileo: the eye, the retina, are mere triggers, though they are triggers with a million teeth. When we are awake, and our eyes are open, they tell the mind what it ought to see—they look outside and then select, out of the varied collection in consciousness's grand gallery, which painting should be shone with light, but they don't do the seeing, no, that's something for the mind alone. For even though the eyes may be shut, as when asleep, or injured, as with a blind painter, the mind still sees, and of its own accord decides what's to be seen. If the images of consciousness are created in the cerebrum, not in the retina, thought Galileo, then of course the lady's eyes could be alive and blind, but the blind man could see.

NOTES

This dialogue between two blind painters, one retinal and one cortical, is based on Gian Paolo Lomazzo and Sofonisba Anguissola. Lomazzo became blind early on (though not necessarily due to methanol poisoning) and turned into a prominent art theoretician. He was a member of the unconventional Academy of the Val di Blenio, dedicated to Bacchus, and used a half-invented dialect to compose grotesque poems. His (attributed) self-portrait as abbot of the Val di Blenio is at the Pinacoteca di Brera, Milan. Lomazzo visited Sofonisba, who became blind at the age of ninety-six, although her blindness was probably of the eye and not of the cortex, as pretended here. Sofonisba's self-portrait is at the Muzeum Zamek, Lancut, Poland. *The Allegory of Sight* by Jan Brueghel the Elder and Rubens is at the Prado, Madrid (apparently modified in some detail). *The Fall of the Rebel Angels* by Rubens is at the Alte Pinakothek, Munich. Another allegory of vision is from the tapestry cycle of *The Lady and the Unicorn* at the Musée du Moyen Age, Paris.

What Frick tells Galileo about the retina and its lack of contribution to vision is quite accurate. Francis Crick and Christof Koch make the point very clear in several publications, especially in Koch's *The Quest for Consciousness* (Roberts, 2004). Complete cortical blindness, notably Anton's syndrome (from the German neurologist who described it as blindness of the soul, or *Seelenblindheit*), is a rare condition resulting from bilateral occipital damage. A similar condition was described by Seneca, in his *Letters to Lucilius* (Liber V, Epistula IX): "You know that Harpastes, my wife's fatuous companion, has remained in my home as an inherited burden . . . This foolish woman has suddenly lost her sight. Incredible as it may appear, what I am going to tell you is true: She does not know she is blind. Therefore, again and again she asks her guardian to take her elsewhere. She claims that my home is dark." Like Sofonisba in this chapter, Harpastes and Anton's patient are blind but deny their condition: they may fall over objects, try to walk through a wall or through a closed door, do not recognize relatives, and instead describe people and objects that are not there at all. These patients indeed have lost the knowledge of what seeing means, but they have a vast store of verbal memories with which they confabulate about "visual" things they cannot even imagine. Such unawareness of deficit (also known as anosognosia) is not unusual with certain cortical lesions. (Other patients may deny that their limbs are paralyzed.) Another form of anosognosia is often found in hemineglect, usually due to lesions in right parietal cortex, in which patients ignore the left side of the world. For example, a painter with hemineglect (described briefly in Chapter 11) might paint his self-portrait with only the right half of the face showing, not to mention dress only his right side, eat from the right side of the plate only, and imagine the right side of things only. For a hemineglect patient, it makes no sense at all to talk about the left side of the world—it simply does not exist, just as for all of us it makes no sense to talk about what the world would look like if we had the brain of a bee.

6

A Brain Locked In

*In which is shown that motor outputs and pathways
are not necessary for consciousness,
nor are they sufficient*

God's flaccid will is ruled by a capricious queen—the circus of disease has endless numbers. Whom would he meet in the next room? thought Galileo. A singer with no voice? A poet without a language?

A man sporting a large mustache was leaning against a cupboard perforated by elaborate woodwork. "No reason to worry about me, Galileo, my gears are spinning tight," he announced.

I trust so, replied Galileo, looking around and wondering where Frick had disappeared to. Then he noticed the brass levers that emerged from the middle of the cupboard, and a shaft holding a thick roll of paper on the left side. He addressed the man: Could you tell me the purpose of this strange cupboard? Is it an organ?

"Oh no," said the mustached man, "though it does belong in a church: the cupboard contains the Automatic Confessor."

A mechanical confessor? inquired Galileo.

"If you want to call it so," answered the man. "It was built by Father P. His pretext? Pilgrims flock to Rome in such large numbers, he claims, that every day friars are taken ill under the strain of too much sin they must remit. So Father P. came up with this machine and sent it here to be evaluated. His mechanical calculator is remarkable enough, but the Automatic Confessor is incomparably more powerful."

How does it work? asked Galileo, who was becoming curious.

"The machine's inner workings are a secret," said the man. "All I know is that there is a set of gears devoted to each sin—one gearbox for *superbia,* one for *accidia,* one for *invidia, luxuria,* and so on. The sinner enters a written confession with these levers, one for each let-

ter, making sure all the details are spelled out, even those he may find irrelevant. At the end, the sinner signifies the strength of his remorse by how strenuously he pulls this red lever on the right. Then, the machine parses the confession, grinds through its secret gears, evaluates every instance of each sin, rates it, weighs it by the intensity of the remorse, and blurts out, on that roll of paper, the penance one deserves. With such machines, says Father P., all the priest has left to do is pronounce the absolution. Of course, a bell will ring if the machine becomes aware of mortal sin."

And has this come to pass? asked Galileo.

"Certainly not," answered the mustached man. "The stench of heresy is so strong that it was smelled in Rome the moment they heard word of the machine. Leaving the absolution to the priest—they saw it at once—is just a fig leaf on the abdication of the soul. If one goes along with this machine today, tomorrow some other machine will replace not just the priest but the Pope, all the saints and angels, and who knows, even God himself."

Maybe, said Galileo, but if it helps a man who is overworked, I don't see anything wrong with it. After all, it is just a machine.

"No," answered the mustached man. "It is a pagan idol that crunches men's offerings in words. And men are fools—they'll soon believe their idol's answers, even if it understands nothing at all. Because it is one thing," he continued, "to count instances of *accidia, invidia, avaritia, luxuria,* and so on, to calculate the proper penance. It is quite another matter to put each sin in its own context. One gear that sifts through the mud of human life and error can only grind its teeth at gluttony, for instance, but it does not know anything else. Another gear sucks in foul sludge peering for lasciviousness, divides the acts and thoughts of men into those that reek of lust and those that don't, but has no idea of what life is, if you take lust away. Lust and not lust, that's all it understands. And even within lust, the many forms that lust can take, the sensuousness of a womblike flower, the enarched, nippled breast of the hill that stretches on the horizon, the obscene beauty of a bird's plumage; even the sublime, unearthly beauty, the resplendent majesty of the virgin, can titillate yearnings not of the sacred kind; even the stirring curve of the mother's voice lamenting the dead son can lead to a frenzied ecstasy the sharpened, ticklish imagination of the hermit."

I don't understand myself, said Galileo, interrupting brusquely.

"I was carried away," apologized the man, taking a little drawing from his pocket. "This is our little church and cemetery. When I sketched it long ago, it seemed innocent enough, but now it's evident what's wrong with it." He pointed at the hill in the background. "But the machine, to know what's wrong with this picture, would need to see like you and me, not just to read, and would have to understand what it sees—the contours of trees and mountains, the shapes of moving clouds, the way that everything relates to everything, in every scene we see. To know of sin the way we do, to uncover it in life, in art, in dreams, it needs to know all that we know, what is, what isn't, and what relates to what and what not. A machine may best a priest in speed and accuracy, will not get tired, and not succumb to empathy, but if it cannot see—the way the true confessor sees, even behind the dark grille of the confessing booth—it will not understand the sinner, nor the sin."

Well, said Galileo, but if it spots the sins better than the priest, on what authority can we say that it understands them less?

"Beware of judging just the acts and not their meaning," said the mustached man. "The doctrine says that *what* is done does matter, but so does *how* it's done. The act may be the same, but it may be reached blindly, or unfold within the light of consciousness."

And how would one tell? asked Galileo, unconvinced.

"Just ask a question, an arbitrary question," said the mustached man. "A good priest can answer any question about what is and isn't. But a machine? See here." And he entered a string of letters with the levers: "How bad a sin is silence?" At which he revolved a large wheel alongside the cupboard, and after some cranking, the paper roll began to flow. On the paper, written in large red characters, was the machine's answer:

"Blasphemy is a sin. False witness is a sin."

"You see," said the mustached man, "it does not see a sin if it is one of omission. The silence of friends—the acts one could have done but did not do, those it cannot judge."

I will try a question, too, said Galileo, if you do not mind pushing the levers. Tell me, asked Galileo, the ambition to do good, is it vice or virtue?

"*Ambition is a sin, doing good a virtue,*" rolled out the machine.

"It's both right and wrong," intervened the mustached man. "Let me put it to the test." So he pulled the levers and entered: "What is more important, the cause or the effect?"

"*The cause is judged by its effects,*" the machine spewed out.

"Don't you think that men should worry about effect, and cause is the prerogative of God?" asked the mustached man.

"*No,*" said the confessor, "*you are not responsible for either.*"

"See," said the mustached man, "this machine could easily fool a lot of people. Most people will think that behind the grille there is an old Father Confessor, one who sees and knows it all. Instead, it's just a forest of gearboxes, each looking for its sin, forgiving of the rest."

Like the cerebellum, thought Galileo. Each gearbox, each module in the machine, is good at what it does. It can answer well and fast within its own domain but cannot see the context.

And yet, he said aloud, at times it's hard to judge who is behind the mask.

"Truer than you think," answered the man. "Sometimes it's hard to judge if there is anybody at all. Which is why I am asking for your help, Galileo, not to divine whether behind the grille is a confessor or rather a machine, but to find out if a soul dwells behind a human face." And he beckoned to Galileo to follow him into the next room.

Within the room the air was dark and heavy. Lying on a bed in the far corner was a man breathing slowly through a hole in his throat. The wheeze that waxed and waned was the wheeze of agony. Galileo moved closer and felt the man's pulse: the heart was racing, but nothing else moved, and the man's body remained stiff and silent. Yet his eyes were open, and he did not seem asleep. Suddenly, Galileo recognized him: it was his old friend M.

Before Galileo could recompose himself, the mustached man spoke in a loud voice. "This is why I need your help, Galileo. M. is no more with us, and he took the formula with him."

As Galileo did not seem to understand, the man hastily explained: "He has found the formula! The formula for the prime numbers, I am sure that must be it. Of course I left for Paris right away, and then, when I arrived, I found this speechless body—as inert as a stone. It has been seven days, without a single sound."

What's the matter with my friend? asked Galileo.

"M. had an accident to the part of his brain that moves the threads of movement," said the mustached man, pointing to M.'s head. "M. breathes, you can hear that, but nothing else moves. Except for his right eyelid, which keeps blinking all the time—some reflex irritation of the eye."

Galileo again took M.'s hand: the pulse was racing faster, but he would not talk. He is alive, thought Galileo, but is he conscious, or, like Copernicus, is he lost in nothingness? How would one know? Then he reflected. What would doctors do to judge whether somebody is conscious? They ask some questions, and if the patient gives a good answer every time, they judge he must be conscious. If there are no answers, they try and make the patient move and look for any sign of purpose, or make threatening gestures and see whether they bring out a response. If they do, perhaps there is somebody there. But M. was immobile, and pricking his skin did not produce the slightest movement, except perhaps the blink.

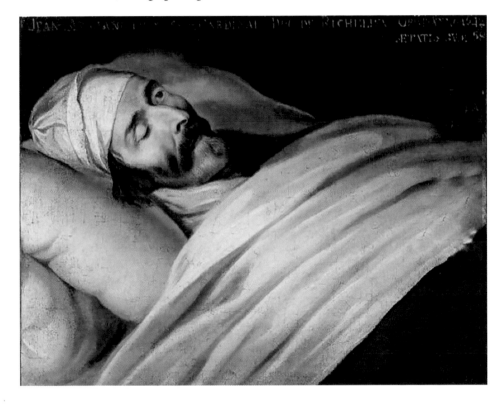

The blink! Galileo was losing faith when it occurred to him. Could it be that M. was trying to say something with his eye?

Could M. hear him? He told M. to open and close his eye to answer, once for no and twice for yes. And immediately, the eye opened and closed twice. Could M. see him? Again, M. blinked twice. Did M. recognize him? M.'s eye said yes.

The mustached man was taken by feverish excitement. In the blink of an eye, he said with childish glee, he would write down the alphabet. And so he did—indeed, he soon began to point to letter after letter, urging M. to blink when his finger was under the one he meant to spell.

Galileo was anxious to discover if M. could feel, and taste, and smell, even more if he could think and remember. He learned it soon enough: M.'s mind was functioning just as it always had, like the well-oiled machine it was. The mustached man instead was trying to pull out from M.'s eye the formula of the primes. But with the symbols of mathematics, the questions were more difficult, and none of them made M. blink.

Meanwhile, M. began to dictate simple words, as if there were no time left. His eye opened and closed so fast, the mustached man gave up transcribing, just moved his finger over the letters and left it to Galileo to read M.'s eye and write all down. "Perhaps the formula is coming out in words, letter by letter," he said. Thus Galileo read what he had scribbled:

> . . . morior morientis mei corporis captivus . . .
> . . . cogito et non ago . . . non ago ergo non sum . . .

"For God's sake, M., tell us the formula instead," said the mustached man. But M. was mute and immobile.

Galileo looked at his friend. M. was proof of this: one can be completely paralyzed and yet be entirely conscious. *"I die a prisoner of my own dying body,"* M. had said. *"I think but do not act, don't act thus don't exist."* So he was conscious, all too conscious.

Galileo sank into his own thoughts. He thought that those parts of the brain that control speech and movement are not necessary for consciousness—they are like ports through which the intentions and decisions of the mind are manifested: all they do is send these orders to the muscles. Of course the orders must be accurate, since they must specify the differences among the words we speak, among the notes we play. Yet M. did not need any of that in order to think, just like the blind painter did not need the nerves leading signals from the eye or the ear in order to see within his mind.

Galileo looked at his friend again. Now M. was blinking furiously:

> . . . spiritus numerus ipse . . .

Then, he said:

. . . No . . . write, Galileo, write now . . . for I have sinned, my
God . . . forgive me, my God . . .

Then, slowly, his eye said:

. . . indulge mihi Domine . . .

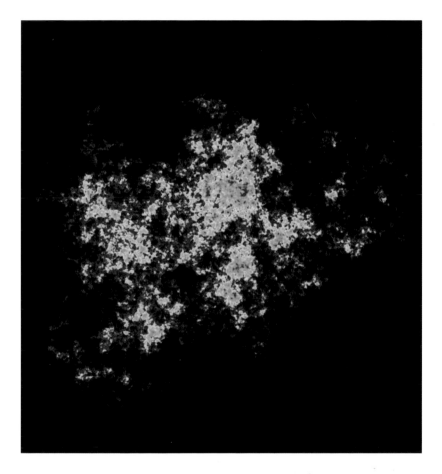

Galileo touched M.'s hand. As when we sleep, he thought, when
we lie paralyzed and yet we are conscious of a dream. But his friend
was paralyzed inside a nightmare—no gentle prodding could awaken
him. Outside a stone, inside a universe of consciousness.

The mustached man came near M., took his other hand, and squeezed it. Then, slowly, he closed his eye.

A soul can have its thoughts and yet hide them forever—unsuspected, undisclosed, unfathomed, and unshared, Galileo thought. But a machine may have no soul and still pass judgment on your sins. One day, perhaps, we will ask not M. but the machine. And perhaps the machine will confess the formula.

Then quickly Galileo left the room and passed by the machine. The paper had rolled down to the floor. He took it and read:

Go in peace, for I absolve thee.

NOTES

This chapter has two goals. The first is to show that one can be conscious even when one is completely paralyzed, like M. because of a brain lesion, or like all of us when we dream. The second goal is to contrast M., who cannot speak or act but certainly has a soul (or rather had one), with a mere machine—the Automatic Confessor, which never misses a beat but presumably has no soul. It seems that at times one ought to be careful before concluding that little response means little consciousness. Or, conversely, that many clever responses necessarily imply much consciousness. What would Descartes have thought of a modern chess-playing program? The mustached man, for his part, thinks that a machine may be good at producing clever responses but bad at producing consciousness because it is made of separate modules. So while it may occasionally fool us with its answers, it lacks the context and understanding that only conscious-ness can provide. His suggestion is discussed in Koch and Tononi, *Scientific American* (2011). "Pascaline," the mechanical calculator invented and manufactured by Blaise Pascal around 1642 (Father P.?), is at the Conservatoire National des Arts et Métiers, Paris. The church tower and breast hill at Pennant Melangell is modified from a photograph by Gerald Morgan. *The Game of Chess* by Sofonisba Anguissola is at the National Museum, Poznań, Poland. It has been manipulated by the addition of the Shadow Hand developed by the Shadow Robot Company in London.

M. is Father Marin Mersenne, a French monk, a man of science, and a friend of Galileo's. Mersenne had tried to find a formula that would represent all prime numbers and had several exchanges on this topic with Pierre de Fermat. After Mersenne's death, letters in his cell were found from dozens of scholars, including Galileo, Fermat, Huygens, and Torricelli. Mersenne thought that the cause of the sciences was the cause of God, and through his immense correspondence, he became a hub for European scholars, who often met in his cell. His chief goal was to promote collaborations to advance science, and he asked in his will that his body be used for research. Mersenne became ill after a visit to his friend Descartes. However, there is no indication that in his final days he may have suffered from complete paralysis. (It is also irritating that the author inserts Latin quotes—if they are quotes.) At any rate, another Frenchman, Jean-Dominique Bauby, suffered from a small stroke in the brainstem that left him completely paralyzed except for one eye. Bauby was able to dictate a short book, *The Diving Bell and the Butterfly,* before dying of heart failure. Yet another Frenchman, Alexandre Dumas, presciently described this condition in *The Count of Monte Cristo.* The locked-in syndrome, as it is now called, leaves patients completely conscious but capable of communicating only by moving an eye up and down. The painting of another Frenchman on his deathbed is by Philippe de Champaigne (at the Institut de France, slightly modified). The colored galaxy is actually a visualization of millions of primes by Adrian Leatherland (www.mysteriousnumbers.com). The final painting is Fuseli's *The Nightmare,* at the Detroit Institute of Arts. During dreams the body is completely paralyzed by brainstem circuits, while consciousness continues. If the paralysis does not occur, some people may find themselves acting out their dreams, with dangerous consequences.

7

EMPRESS WITHOUT MEMORY

*In which is shown that many brain circuits
that help us see, hear, remember, speak, and act
are not necessary for consciousness*

There are powerful machines inside the brain, and consciousness is wise to use them all. Some are for reading the features of the world, others for acting upon it. Most delicate are machines that calculate, and plan and counsel, and then machines that store the memory of events—though they are all just tools, not parts of consciousness itself.

In front of Galileo was no machine—in front of him stood the lady who had bewitched his soul, when he was young, and lured him many times to Venice. He said to her:

> *Among the fair your learning is renowned,*
> *Among the learned the glory of your fair looks,*
> *And both you conquer in study and beauty alike.*

In mellow voice, with studied poise, she answered:

"A charming man you are, who with sweet verse announces his desire. But be at ease: it is my sacred duty to attend to every poet."

Milady, you know the verse is yours, and yet with noble modesty you feign indifference to praise.

"Dear Prince, for sure you must be one, pray tell me who you are, and whence you come to me."

I must have grown wizened and weary, milady, if you don't recognize your humble suitor. Should I concede again that science should bow to poetry, and present arms? Give me your secret watchword, and I shall tell you mine.

"Dear Prince, the gift of memory was stolen from my jewel box. One day I woke and I had lost my past. I live a recluse of the present, and in my house there is only one room."

That cannot be, milady, your memory was the marvel of Venice. Of poems and people alike, you knew by heart every secret sense.

"Not anymore, Prince. Now every poem is singular to me—each poem gives me new joy—though you will say I heard it many times. But love is pure when love is fresh, unburdened by the dust of memory. So a virgin pleasure I receive from all, from all who deign to visit my abode—each time they shall have me for the first time: no dullness will set in, no weary custom—each time anew discovering my Prince, a trembling bride on her first night."

How does it feel, milady, not to remember, to hear the burning words of passion of your friend, to feel the mighty impression of his wish, and yet they leave no trace?

"I am not sure," she said. "I argue a little with myself." And then she said: "Perhaps it is like this: like waking from a dream every time I wonder. Alas, a dream from which I wake each moment of my life—each time I feel I am alone, every instant passes without a past. Floating amidst the river Lethe I can't behold its shores, and when I do come up to breathe, nothing for me to grasp; when I look back, nothing for me to see."

Then tell me this, milady: The heat of love, is it as ardent as it was before, the bite of pain, or pleasure's beauty—are they as intense, when they are naked, as when they dressed in memory's wearied garments?

"Do not importune her anymore," said Frick, who had suddenly reappeared, taking Galileo away. "Indeed her memory of many events has vanished, and before long you would have made her cry. Something happened to her brain—deep in her temporal lobes, the hippocampus must have been destroyed. It is a thoroughfare that collects, at the end of tighter and tighter funnels of nerve fibers, the separate strands that make up every event of which we are conscious, one strand from each region of the cortex. And by collecting those strands in a single place, the hippocampus can weave them quickly together, before they are dispersed, so that we can recall the whole event at once."

Frick went on: "You have just seen your lady in a remote corner of the hospital, inside a room you never saw before. The hippocampus will weave a knot between that room and her lovely face and keep it tightly in store. So if one day you see that room again, the room will pull its strand in the hippocampus; that single strand, woven together with that of her fair image, will pull her up as well. Then through an inverted set of funneled strands, fanning out to reach the full width of the brain, the nodes in the hippocampus will call back their sources in the cortex and recall what was going on when that memory was formed. But now the thread of all the past is lost to her."

Galileo would not forget his lady's eyes. So he turned back, addressing her again:

Milady, confess me this: that your forgetfulness is the secret of your enduring beauty, that with the gods you have struck a pact—lest time would leave its mark on your fair body, it would not leave a mark on your swift mind.

At which she said: "I am not sure, for I am arguing a little with myself." And then she added:

"Dear Prince, for sure you must be one, pray tell me who you are, and whence you come to me."

So Galileo let Frick lead him away.

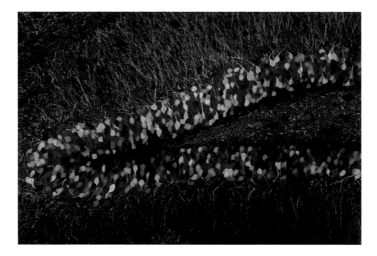

His thought was wandering along the hippocampus: Why was it so, he wondered, that despite loops made of a million strands, which tie the hippocampus with the cortex, the hippocampus does not partake of consciousness?

"The hippocampus is only a slave, one among the many slaves of consciousness—what it can offer her is the power to remember and recall," answered Frick. "Without her slaves the empress has no clothes, is chained to an everlasting present—she sees, she hears, she feels and thinks, but cannot remember any event, nor can she imagine novel ones: imagination is the twin sister of memory. No, despite all those connections going both ways, the hippocampus does not partake of consciousness." Then Frick added: "This is just one example; many other loops leave the cortex, descend into the lower portions of the brain, attach to pegs and pulleys there, and return to the cortex. These wires, and the pegs and pulleys to which they are connected, are the bookkeepers of the cortex, and run its skilled abacuses. They do the reckoning that the cortex needs, but are not lit themselves by the light of consciousness. Indeed, even within the cortex, there must be many more such loops, which digest images and sounds, parse objects and words, and weave actions and sentences together."

Galileo's expression was not one of persuasion, so Frick asked him: "Tell me then, who invented the funnel for prime numbers?"

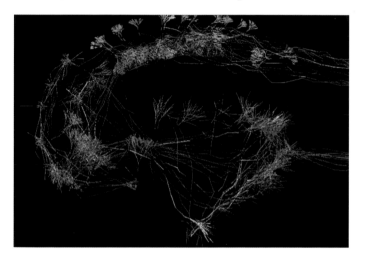

Galileo hesitated just briefly, and then his lips proffered the answer: I know: the sieve of Eratosthenes.

"How did you retrieve the sieve and the name?" asked Frick. "Think of it. You certainly heard the question in your consciousness, and then? And then consciousness pulled some strings in your brain, and lo and behold, the answer came to you, without you knowing at all what was going on: there it was, Eratosthenes, on a golden plate. These wires that loop from the top of the brain to the ganglia below, and then back to the brain, are the slaves of the mind. They promptly perform their duty, though they don't know what they do, nor why, nor for whom: they are blind, dumb, and mechanical, and there are millions of them, as in the cerebellum. Without them, life would be an immense hurdle—days would be spent tying your shoes and buttoning your jacket." And he guided Galileo into the next room.

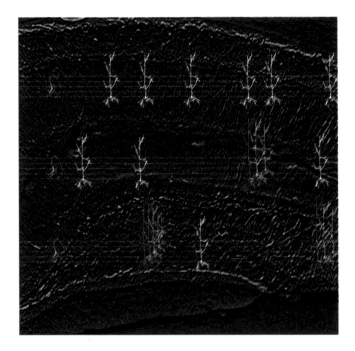

There it was dark and quiet. But then the silence resounded of a single, low-pitched note, the note vibrating full and rich, etched out of stillness in great relief, lasting suspended across time, until it slowly disappeared into dead air—the sound of a viol string.

Time passed, and stillness reigned again. Then another note, lower-pitched and dark, reverberated through the room. Two breaths, and then another note, and then a pause that would not end.

What music is this? Why is the playing so slow? Galileo finally dared to ask. Frick whispered that the player was ill, every act was slow and deliberate: "What once was automatic, without effort or consideration, now requires conscious thought, every act needs thought. No more scales rushing through the fingers, but only single notes, every single note thought out in depth, decided as if it were the only one that can fill the compass of the present."

Then a voice spoke, slowly: "Only the essential note is ever played—every note in solitude, its whole life heard—with time enough to extinguish into silence."

A candle was lit, a woman's shape emerged from the shadow, and Galileo knew who she was.

Time had caught up with all his muses—their suitor soon would be the next in line. Already hand and thought had lost their nimble leap—one day not even memory would stay behind to mourn.

NOTES

The verse is from the Venetian poetess and renowned courtesan Veronica Franco (133–36, I, *terze rime: e così 'l vanto avete tra le belle / di dotta, e tra le dotte di bellezza, / e d'ambo superate e queste e quelle* . . .). The presumed portrait of Veronica is attributed to Tintoretto (Worcester Art Museum, Worcester, Massachusetts). Galileo indeed used to visit Venice during weekends when he was a professor in Padua, in order to have a good time away from colleagues and students. The photomicrograph of a section through a "brainbow" transgenic mouse hippocampus is by Tamily Weissman. The reconstruction of hippocampal circuitry is by Giorgio Ascoli; the picture of neurons superimposed upon the hippocampal score is courtesy of Gyuri Buzsaki. The plight of consciousness without memory was first revealed by the case of Henry M., a man whose hippocampi and neighboring structures were surgically removed in an attempt to relieve his seizures. His story is movingly told in *Memory's Ghost,* by Philip Hilts (1995, Simon & Schuster). Henry did in fact often say, "I'm having a little argument with myself." One should perhaps be more careful than Frick and Galileo in dismissing any direct contribution of the hippocampus to consciousness, but their point is plausible. The portrait of Barbara Strozzi is by Bernardo Strozzi (Gemäldegalerie, Dresden). Barbara, born in Venice of a family connected to the Medicis, was one of the greatest musicians of her time and also, perhaps, a courtesan. Here she exemplifies the usefulness of "unconscious" modules to perform actions that can be made automatic.

8

A Brain Split

*In which is shown that consciousness is divided
if the brain is split*

*Two minds are housed inside one head,
each musing its own destiny.
Each day they bond to sing in tune,
but strife can break their harmony.
One song turns two—their tune is shred.*

And so it was, a choir of servants chanted, standing around a man in mourning, and at each beat a boy flayed him with a violin bow. With clenched teeth, the man turned and spoke to Galileo: "I called for you to show you my discoveries, so you can testify that I am not out of my mind." It was the wild Prince of Venosa.

"The first discovery was long ago," said the Prince, "it was one of my servants, whom I had taught to play at a young age. His right hand was a marvel, the nimblest hand that ever touched the clavichord. But his left hand! Fie! That hand was like a hoof, as if a brute had struck the keys—as if an angel and an animal were housed in the same body. So when the time had come, I asked my doctor to find out, by opening the skull. And when Salerno did as I had told him, we saw his brain, and this is what we saw: the angelic mind that drove the masterly right hand dwelled to the left, not to the right, as had been thought before."

The Prince raised a large glass vase. Inside was the servant's skull, the brain preserved intact, but in the skull was only the left hemisphere. "On the right," said the Prince, "was emptiness and negation—the emptiness that marks the devil."

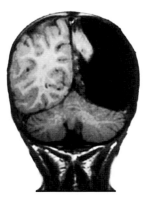

    Devil or not, thought Galileo, that rather proved that half a brain was good enough for one full mind—a mind that could both speak and play and do the multitudinous things a mind can do. But Galileo could not think further. With a wink, the Prince released the servants and led him down a dark spiral staircase.

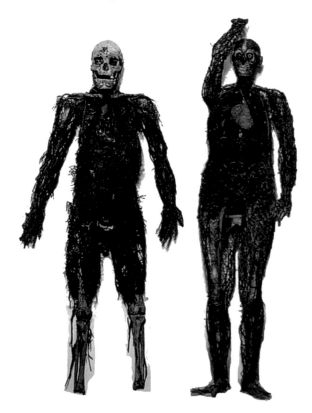

Inside the crypt, a recess in the wall held two scarnified bodies, a man and a woman. In the middle was a marble table, and there, shrouded by a veil that also seemed of marble, was a young man, his arms outstretched high over his head, his skull wide open. Salerno was behind it—spidery and dark—handling two copper grids made of tiny tubes, which were connected to two large vats nearby: one full of ice, the other full of boiling water.

"The servant could not keep his eyes perfectly still," whispered Salerno, "so I applied a paste of my own making. Since then he has been staring straight ahead."

"This is my most remarkable discovery," said the Prince to Galileo. "The servant whom you see here rest, can use both hands with perfect independence, and has done so since he was a child. But halfway through every piece he played, perhaps because the devil is envious of perfection, he stiffened, screamed, and foamed, shaken by a seizure. And thus his skills have come to nothing.

"So at my orders," the Prince continued, "Salerno opened his skull, to see whether the left and the right brain were connected, as they are in all corpses, by thick callous fibers straddling across the midline, or they were separated, like two musicians, man and wife, who severed all relationships—but they still live and sing in the same house.

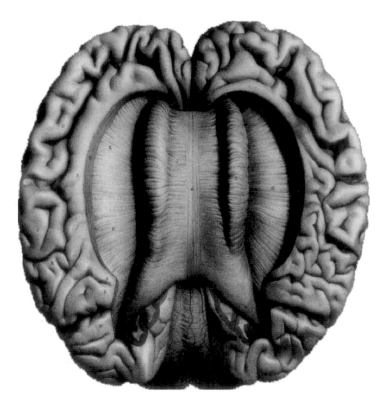

"Salerno is a magician," said the Prince. "He opened the skull alive and without causing pain. For him I crafted a covering of Venetian crystal to protect the brain, to watch it work when it will play.

"And yet we were surprised," said the Prince. "The two hemispheres were not estranged, but connected, just as they usually are. Salerno then constructed these capillaries of copper, which can be turned cold or warm in a few instants, by flushing them with water that is icy or hot."

"If you freeze the whole surface of the brain, Signor Galileo," said

Salerno, "all the animal spirits disappear—he turns blind and deaf, mute and paralyzed, just like a corpse. But if you freeze the right brain only, as I am doing now, you'll see his left arm fall, flaccid, as if inert and devoid of life." Indeed, the arm fell down and lay to rest on the marble veil at the servant's side. "While his right brain is stunned by the cold, his left brain is alive, and his right hand can play the most moving of songs."

"Tell the visitor who you are," the Prince ordered the servant.

"I am Ishma—the desolated one," he said as if under a spell.

"Now watch," exclaimed the Prince, while pulling a rope to unveil a large marble complex, which stood in front of Ishma's staring eyes.

"Tell what you see, Ishma," said the Prince.

"I see the body of a nymph, radiant of beauty, but veiled with modesty, her face is sweet to me."

"Do you see anything else?" asked the Prince.

"No," said Ishma, "she should be crowned with lilies. The garden of Eden should embrace her, and birds should sing her praise."

"Do you feel unwell, Ishma?"

"No," replied Ishma, "nothing hurts me, but I can't move my eyes."

"You'll be able to move them soon enough," said Salerno—and

then, addressing Galileo: "In this way, with the eyes straight ahead, his left brain only sees what is right of the middle. And his right brain sees just what's on the left side. So we can ask each brain intimate, private questions, without letting the other one know."

It's clear, thought Galileo, the left hemisphere is quite enough to uphold one consciousness, a consciousness that sees and hears, that thinks and speaks. You need two legs to walk, one hemisphere is enough to think. And yet something was missing: it was a consciousness that only saw the right side of the sculpture.

"It's time to switch," said the Prince. "Do so, Salerno," he ordered.

Salerno moved the icy copper grid over the left brain and applied the warm grid over the right brain, so that it could recover its spirits.

"His right brain does not speak as well," the Prince warned Galileo. "Say who you are," ordered the Prince after a while, when he saw that the servant had raised the left arm and lowered the right.

"El," said the servant's right brain, and then, after a long pause, "God himself."

"Did anything happen to you, El? Do you feel strange?"

"No," said El, again after a pause.

"Have you changed?"

"No," said El.

"And what do you see in front of you?"

It took some time, but then El, hesitantly, said:

"Man beast."

"What is next to the man?"

"Nothing," said El.

"Do you see anything else?" asked the Prince.

"Man beast," said El.

The right hemisphere too, then, thought Galileo, has all it takes to uphold one consciousness, a consciousness that sees and hears, thinks and speaks, though it speaks slowly. But this time, it was a consciousness that saw only the left side.

"Now watch again," said the Prince. "Heat both his brains, Salerno, and do make sure the callous fibers in between are warm."

And when Salerno did as he had been told, both arms were raised as if in prayer. The eyes, however, remained closed.

"Who are you?" asked the Prince.

"Call me Ishmael," said the servant. "I am the one who can hear God."

"Open your eyes," ordered the Prince.

The servant let out a frightened scream.

"Why do you scream, Ishmael?" said the Prince imperiously. "What is it that you see?"

"I see them both, Prince, I see them both together. I see my lady fair, I see the hoofed swine, the one who mounted her in beastly embraces. I see they twine together, and I am frightened that their reward is coming."

The Prince touched the servant's eyes with a handkerchief, as if to soothe him.

"Reward has come, Ishmael. Your lady now is pure. And I have cleansed her, so her poor body could be left by her pure soul."

The Prince was still as mad as ever, thought Galileo, and playing games with his servant's mind, or rather his two minds. The game was interesting, however. The left brain was Ishma, the right one El. Yet when Salerno warmed both brains, he did not conjure up both of them separately. Instead, Ishma and El became one consciousness, became one Ishmael, and Ishmael could see the entire statue.

What happens if you cool not the brain itself, but just the fibers that connect the two hemispheres? asked Galileo.

Salerno was flattered. "An astute question. When I did so," he replied, "both hands did keep their strength. But Ishmael was gone. Instead, he split into Ishma and El, one conscious of the right side of things, the other of the left side."

So consciousness is split when the brain is split, thought Galileo.

Then an image came to him: he saw the servant's arms, the left one raising and the right one dropping, then the other way around, raising and dropping at faster and faster speed. Tell me—Galileo asked Salerno—when you freeze his right brain and warm it, then freeze it and warm it again, and you do it fast, every swing of the pendulum, what does it feel like? Does he feel that he is changing, or does he think that he remains the same?

"Each mind, each brain, will soon get used to its prison, not knowing what it's missing," said Salerno. "You cannot see behind your back, but nothing seems amiss to you; though if you were a pigeon, surely you would feel half blind. So a man who's blind from birth will never know what seeing is like—will never know what's missing—and if he had been raised alone, he'll think that nothing much is wrong, but like a mole burrow his days away unseeing and unaware. Once I was told," continued Salerno, "that savage people in America possess another sense, a sense besides hearing and sight, and touch and taste and smell—a sense through which they feel within their heads, with startling strength, whether one's words and acts are truthful or are not. Like we can sniff if wine is good or bad, they sense both truth and falsity inside their minds. And when they sense a vivid lie, they crumple to the ground, holding their growling heads. But have you ever felt it missing, this strange sixth sense of truth? Does your own mind feel robbed, because it does not have it, and sense an empty urn between your eyes?"

How would I know? I'll never know all that I'm missing, said Galileo. A mind's a universe, indeed, but it's a universe with bounds. One cannot step outside its borders, it does not matter how large it is inside.

"Yes," said Salerno. "So how much mind does a man need?"

NOTES

Carlo Gesualdo, Prince of Venosa, almost the same age as Galileo, was noted for his peculiar style, not just in musical matters. He discovered his wife and her lover in flagrante delicto. After butchering them to death with his sword, he had their bodies hanged in public from his palace. Later he suffered from depressive bouts, and his young attendants were required to flay him mercilessly. His palace in Naples passed to Raimondo de Sangro, Prince of Sansevero, a man devoted, like his predecessor, to the singing of young boys and to alchemy. Legend has it that he and his physician Salerno injected a man and a pregnant woman with a secret petrifying substance to obtain the perfect casts of their blood vessels and organs that are kept in the palace's chapel. Just as secret is the procedure by which Salerno and Sanmartino were able to amalgamate a marbled veil onto the body of Christ, displayed at the center of the chapel. The next statue, also in the chapel, is the *Pudicizia* (*Modesty,* by Corradini), based on the Prince's mother: its allegorical meaning was that, to possess knowledge, one must lift her veil. The next statue, the *Disinganno* (*Disillusion,* by Queirolo), represents the Prince's father, trying to free himself from false beliefs with the aid of the intellect. The picture of the cerebral hemispheres, split open to reveal the corpus callosum connecting them, is from the atlas of Achille-Louis Foville's *Traité complet* (1844). There are indeed many patients who had to undergo a hemispherectomy and have led a near-normal life afterward. The experiment performed by Salerno is inspired, with some liberties, by the so-called Wada test, in which an anesthetic is injected first into the circulation of the right brain, and then of the left, to find out what are the independent functions of each hemisphere. Usually the left hemisphere is the one who speaks, or speaks much better. Classic experiments by Gazzaniga, Bogen, and Sperry demonstrated that when the brain is split, each hemisphere sees only one half of the visual field, and the other one does not know about it.

9

A BRAIN CONFLICTED

*In which is said that consciousness can split
if different regions of the brain refuse to talk to each other*

*"Around my heart three men have come, seating outside, not in where
love abides, the love that sole is lord upon my life."*

The woman saying these words advanced with her hands outstretched. A man named Piso was holding her arm and directing her
gently.

Another blind lady, thought Galileo. The way she dressed, she
might have been a nun. What is the matter this time? he asked, turning to Frick.

"This lady is quite intriguing," answered Frick. "Unlike the two blind painters—the man without the eyes and the old dame without the visual cortex—this lady has nothing missing. Indeed, she can actually see, except she sees unconsciously. Watch," said Frick, pointing a long stick toward the advancing woman. But at the last moment, she veered to the right. Again Frick pointed the stick at her. And again, at the last moment, she turned and walked away.

"Why did you turn just now?" he asked the woman.

"I thought I heard a noise" was the reply.

"You just witnessed it," exclaimed Frick. "She sees, but she denies it. I am convinced this is what happened to her brain: for some reason the coalition of neurons that gives rise to consciousness decided to retreat from the visual parts of her brain. Nonetheless, some visual modules there are still at work, but they do so unconsciously—like zombies."

What is a zombie? asked Galileo.

"A mechanism in the brain that does its job unconsciously, like an automaton," answered Frick. "In Teresa's case, a small, dedicated machine inside her brain that can detect an obstacle and steer away from it, but does so independently of Teresa, who is unaware of it."

"Well put," intervened Piso. "Teresa's symptoms come from the brain—I have said it for years—but my colleagues still think it all comes from the womb, and force her to wear medicated pessaries and fumigate her insides."

Then Piso whispered to Frick and Galileo: "She is no ordinary lady—Teresa can read her Latin, write her poetry, and through her wit and logic she put to shame well-seasoned scholars. But that was also how her misfortunes started. She fell in love with a philosopher—a young ambitious squire who used his intellect like a knife, with which he carved to pieces his opponents. And soon enough they paid him back by carving out a piece of him. To teach a lesson to the lovers, they plucked him right in front of her. That day Teresa became ill: for seven months she could not see, and even now she often faints, at times she cannot stand, and other times she becomes blind."

What happened after that? asked Galileo.

"The usual," answered Piso. "She went to a nunnery. Teresa, will you tell us what happened to your squire?"

Teresa lowered her eyes. "We once were wed in secret, but he would not be parted from his worldly spouse, not even for a day—philosophy, he said, cannot bear rivals. To me the name of wife was nothing: it's sweeter to be a mistress—love and be loved—so that our faith and love must be professed anew and with fresh will every single day, not forced on two tired prisoners by the chains of marriage. So when they parted us, I made my will his own, and to be faithful to his love, I married the only one whose vows increase one's virtue. Yes, I long to know where he may be. And yet I fear that since I took my vows, his love has wilted, for he has never called for me. I fear I would forgive, and to forgive redeems the soul, but love is doomed."

"Do you remember how you and he were parted?" asked Piso.

"One night I fell asleep with him, I heard a scream, and then it was all dark. When I woke up, I felt as if my heart were taken from my bosom. And now it's beating elsewhere, and I am split in two."

"She may know better than she thinks," said Piso in a low voice. "I am not sure the part of her that sees is truly unconscious. All while she insists she is blind, her left hand can do things no blind person could do. Let me show you," he said, and holding Teresa by the hand, he helped her to sit, while he picked up a deck of cards.

"Tell me, Teresa," he said, "do you see these cards?" And he spread them out on the table.

"How can I see them if I am blind?" she answered.

"Which card is this?" asked Piso, picking a four of clubs.

"How would I know?" replied Teresa.

"Which of these cards do you like best?" asked Piso, spreading the deck so that the Jack of Spades stood out right in the middle.

Teresa blushed visibly. "Doctor, you are teasing me. Are you playing some tricks?" And while saying so, Teresa's left hand reached for the Jack of Spades and deftly turned it facedown on the table.

"Isn't it remarkable," said Piso, looking at Frick and Galileo. "A part of her brain—the one controlling the left hand—can see the cards perfectly well—and seems to be much smarter than a zombie."

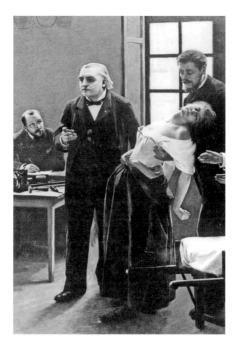

How do we know that she is not feigning her blindness? quipped Galileo.

"Why would she feign it?" said Piso unperturbed. "A few times I took my time to observe Teresa when she thought she was unseen, and still she acted blind. Besides, sometimes she is not just blind but in a trance. Then one can squeeze her hard above the clavicle, or prick her with a sharp nail, and she does not complain."

"Quite interesting," said Frick, looking at Galileo. "Teresa has hysterical blindness, or so they used to call it, thinking it was due to vapors from the womb. But what we saw just now makes me agree with Piso. If the part of her that sees can understand which card is which, blush at the Jack of Spades, and act appropriately in many situations, it's hard to argue that it is unconscious. And so there may be a second, minor consciousness inside her brain, one split from Teresa's dominant, major consciousness—the one that talks to us."

Galileo thought of Ishmael: his consciousness would split in two, into Ishma and El, when Salerno split his brain by freezing the nerve fibers connecting the two sides. Did Teresa's consciousness also split in two, between one part that spoke and one that saw? And the Teresa that saw but could not speak, was she like El? A clandestine passenger hiding inside the skull, coming to the fore only if unseen?

"That's just what I think," said Piso. "When Teresa has one of her spells, it is as if a blind woman and a mute woman were housed in the same skull, not knowing they live together. Perhaps a transient gap has risen inside her brain and separates the part that speaks from the one that sees."

"Indeed," approved Frick, "perhaps her brain is occupied by two coalitions, struggling for control: they fight over the cortical territories, advance and retreat, split and merge—right now they are split. The one that hears has taken control of the brain's northern territories, she speaks to us, and decides what to do next. The one that sees has been excluded from those dominions and reigns over the south, the visual regions. There she sees and understands but cannot speak, and only occasionally she manages a sortie—a swift incursion done with the left hand."

"Just so," commented Piso. "And bear in mind that the splitting and merging of her mind can happen in an instant. Let me try something now—I just received a package that should please her." He winked at Frick and Galileo, and then addressed Teresa:

"Teresa, you have received a present, can you see it?" said Piso, placing a small wooden box on the table. Teresa shook her head. "It comes from somebody who greatly admires you and wishes to regain your confidence." Teresa shook her head again.

Piso opened the wooden box, full of tiny gears and bells. He turned a crank, and right away the box began to sing, chirping like a bird.

"Lo and behold!" said Piso to Frick and Galileo. "A music box! And what a song! On summer nights in Paris, young men would sing it to their beloved. And best of all, the author of both text and melody is . . ." But Piso stopped, looking at Teresa.

Her arms outstretched, she had fallen to her knees and closed her eyes. Then, almost imperceptibly, Teresa began to sing along, swaying her head softly. Then, when the song was over, she raised herself and quietly cried. "My song!" she murmured. "My song—his song—the song that says he thinks of me all the time. The song that breaks his silence and stirs my joy." Slowly, Teresa opened her eyes, looking toward the sky, and then she cried again: "My Lord, I am one again, and see, see the soul's fountain in a bed of crystal, the divine sun, its brightness dazzling, the golden spear ending in a point of fire."

Piso gently took Teresa's hand. "I knew hearing his words might heal you, though they may just be a melody in a box. You see, gentlemen," he said, addressing Frick and Galileo, "it is as if, now that Teresa heard her beloved song, the visual parts of her brain were reconciled with the speaking parts; as if a partition were removed that kept wine and water separate, and the two were allowed to mix again, and become one."

Or maybe, thought Galileo, it was as if one were conversing with a friend in a gusty wind. When the wind is blowing, your words do not reach your friend, nor his you. Then suddenly, between one gust and the other, the conversation can proceed. Maybe the parts of her brain were separated by some nervous wind, which blew when she was distressed.

And maybe such a wind did not just blow inside Teresa's mind—maybe inside his own head, too, clandestine portions of his brain were hiding and plotting their furtive subterfuges, laboring day and night to pursue their hidden schemes, doing things he would not want to do, perhaps turning him sad or gay, friendly or ill disposed, and would not tell him why; nor could he speak to them, and let them hear the voice of reason. How many twins were there who shared his brain? How many selves were housed within one's mind?

NOTES

In 1618 the French physician Piso (Charles Le Pois), trusting, as he said, to experience and reason, overthrew the doctrine of hysteria that had ruled for two thousand years, showed that it occurred at all ages and in both sexes, and that its seat was not in the womb but in the brain. It is possible, though by no means proven, that certain symptoms of hysteria (now called conversion disorder) may be due to the functional, reversible disconnection among brain regions, and to the blockade of some crucial pathways. Teresa's mind surely is split, in no small measure because she is herself a medley of multiple personalities. In part she is modeled after Héloïse, the lover (and secret wife) of Peter Abelard, who after suffering the notorious assault, perhaps instigated by rival factions, forced her to become a nun. He was not only a philosopher but a composer of poems and songs, the early ones love songs addressed to her; the later ones, when he somehow lost interest and had become a monk, turned to sacred topics. One of the few melodies of Abelard that passed down to posterity comes from a hymnal and bears the strange title *"O quanta qualia."* Another source is clearly Saint Teresa of Ávila, who is paraphrased when sight is regained. As far as one can tell, neither Héloïse nor Saint Teresa was affected by hysteria, although Teresa's ecstasy and other peculiar religious phenomena have been considered at times as bordering on the pathological. Hysterical phenomena, such as paralysis, blindness, fainting, and aphasia, can also occur in acute instances of romantic love, though this part of the literature is curiously neglected and deserves further study. Finally, and most to the point, Teresa is modeled after several patients described by Pierre Janet, who in his medical thesis, *L'état mental des hystériques* (1892), introduced the concept of "dissociation" between psychic functions, emphasized the link between symptoms and early trauma, and gave many examples of the role of the "subconscious."

In this regard, the question Frick and Piso briefly entertain is a difficult one: Are "unconscious" perceptions and actions truly unconscious, carried out by functionally isolated modules, also called "zombie systems" (Crick and Koch, "A Framework for Consciousness," *Nature Neuroscience,* 2003)? Or are they conscious on their own, like the nondominant hemisphere in split-brain patients, just hidden away from the dominant consciousness? And how would one know? At the end of the chapter, Galileo ponders how much he is really in control of his own brain and actions. Is there a hidden Galileo inside his skull, a shy, clandestine one, who may see and do things the dominant one may prefer to ignore, hold unsavory prejudices, and cause slips of the tongue? But who is really in charge? And what does this clandestine think, being carried along without much say? The initial reference is to Dante, *Rime,* XLVII. The eyes are from *The Ecstasy of Saint Mar-*

garet of Cortona by Giovanni Lanfranco (Palazzo Pitti, Florence). The painting by André Brouillet shows Charcot at the Salpêtrière in Paris, demonstrating a hysterical fit of Blanche (Marie Wittman). Blanche is supported by Charcot's student the neurologist Babinski. Janet and Freud also attended Charcot's demonstrations. The photograph of a patient is from Charcot's *Iconographie photographique de la Salpêtrière* (*Attitudes passionnelles: exstase, 1877–80*). *Abelard Soliciting the Hand of Héloise,* by Angelica Kauffmann, is at the Burghley House Collection, Lincolnshire. *The Bird Organ* (*La Serinette,* or *The Music Box*) by Jean-Baptiste-Siméon Chardin is at the Louvre, Paris. Bernini's statue *The Ecstasy of Saint Teresa* is at Santa Maria della Vittoria, Rome.

10

A Brain Possessed

*In which is shown that when cortical neurons
fire strongly and synchronously, as during certain seizures,
consciousness fades*

In the broken light of the lanterns, Galileo could make out a narrow courtyard, surrounded by three or four rows of planks. The planks were creaking and shaking, crowded with shabbily dressed villagers, leaning out as far as they could, and shouting in a language he did not understand. The shouting was waxing and waning unpredictably, as if conducted from high above. The women were holding babies in their arms. This land is plagued by seasons of foul weather, thought Galileo, and covered his ears with wool as gusts of wind blew and froze the night.

In the middle of the courtyard a girl—she was maybe sixteen—was tightened to a wooden wheel, a rope around her neck and one around her waist. On one side her long hair was sinking into a filthy bucket. Galileo took hold of his telescope to see her better: a fat friar was busy over the girl, his face invisible behind a black hood. Next to the friar was a man who looked like a civil servant—he was reading aloud, it seemed, but his words could not be heard.

Somebody was pulling at Galileo. Two children with pinched, old-looking faces were holding him by his coat. He threatened to shake them off, but they would rather be beaten than let him go. With a sneer the children handed him a note written by a firm and polished hand, but they would not let go of the note and were tearing the paper apart. Galileo had to twist their wrists a full turn and with all his force, but their expressions revealed no pain. Then one, still sneering, said to the other, "One of us, one of us," or so it sounded to Galileo.

The note was signed by the most learned doctor Helmstadt. It said the girl was epileptic, and hopefully she would be the one, at last, to prove the doctor's theory correct. The doctor's ideas were expressed in a roundabout manner: he seemed to claim that seizures were due not to any supernatural cause but to the vital spirits in the brain, when they were active in excess. The villagers and boorish friars may well think her a witch, who during her convulsions had intercourse with the devil, but he knew better. "I count religion but a childish toy, and hold there is no sin but ignorance," the doctor had written.

The fat friar was shouting in vulgar Latin, his head facing toward the sky: now with the help of God, he would obtain a confession. He sent away a woman who had crossed the court in a fury, a large pan in her hands: yes, all the beans had rotted, it surely was the witch's doing, but there were graver charges. The civil servant began to turn the wheel slowly, as if to make sure that no screech would be missed. Now Galileo could hear him speak in good Latin. "A beautiful woman is like a gold ring in the nose of the sow," he said. He wore a hat turned up at the sides. The doctor, thought Galileo, must be the man who was approaching the girl from the rear with some tool in his hand.

The girl could not be heard—the crowd was bellowing again. From behind, the doctor touched her head with his hand, as if to reassure her that everything was as it should be, and gently stroked her hair, as if to judge the quality of a precious fabric. Then, delicately, he probed her skin with the tool he had at hand—clearly he was trying to cause no damage. But then suddenly he raised his hand up high, holding what now seemed like a large scalpel: in a few savage strokes, he tore off a handful of hair, still attached to the bone underneath, and went around the opening, then seemed to wave to Galileo. With an oil lamp, the doctor threw light on the exposed surface of the brain and quickly wiped it clean with his own fingers. The brain was pale like the girl's skin.

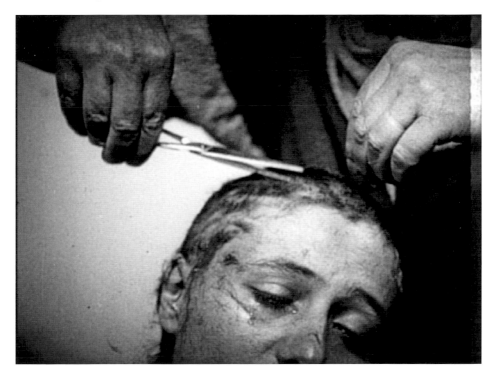

Galileo would have wanted to yell, but the din of the crowd made it hopeless to be heard. And he would have drawn attention to himself, which was unwise, as one never knows what an unruly mob might do. The two children, too, were an impediment. Without him realizing it, they had drawn him much closer.

The doctor was talking to the girl and trying to calm her. The friar lifted his arms, and the crowd turned silent. Staring at the opening on her head, the doctor motioned to Galileo. A church bell rang, and through his telescope Galileo could see a sudden swelling of the brain. The doctor's eyes were beaming. He asked the girl whether she wanted to say her prayers. She wanted to, and when she did so, the swelling went up and then went down again. Praying made her brain flush, thought Galileo.

The men were arguing. The doctor promised that it would not take much longer. The civil servant was impatient. Schleppfuss—this was his name—had been appointed judge at his young age because of his good services, or so he said. His guiding principles were simple: every sausage has two ends, and guilt is never to be doubted. He was glad to serve the interest of science, so far as science would remember it must serve a higher interest.

The doctor offered the girl something to drink, and she took the cup avidly. Suddenly she gave a shrill cry, and a moment later her limbs stiffened violently against the wheel, her back arched against the ropes. The doctor was asking questions, was touching here and there, but she would not respond. "She has lost consciousness," he said. "One sip has been enough to make her go," he said, "just as I thought."

Galileo looked at her face: her lips had turned blue, her eyes had rolled upward, her teeth were gnashing. The girl now foamed at the mouth, the foam turned red, and her body began shaking wildly.

The doctor's eyes were flashing: under the lantern, the surface of the brain had been flooded with blood. Finally, he had obtained proof: the seizures were due to her brain swelling with excitement. The friar's teeth, too, were flashing under the black hood. He too had received proof—she was possessed—the devil was tossing her around. And then her body fell as a dead body falls, and snow was falling on her ashen skin.

"In a little while she will recover," said the doctor, who was suddenly behind Galileo. He was a middle-sized man, wrapped in a black cape that closed at the throat with a little metal chain, his face still smooth, but his cheeks already flabby with age. "They think she has been seized by evil spirits, the uncouth rabble. Of course there are no evil spirits. It's just the animal spirits, those that make her nerves quiver. The girl has too much sulfur in her brain, so her animal spirits burst into excessive discharges. That causes her body to stiffen and jerk."

What was it that you gave her to drink? asked Galileo.

"Camphor oil," answered the doctor with a smile. "She was epileptic, to be sure, but I had to make certain she would have her seizure at the right time."

Galileo thought of the books Frick had given him: seizures are indeed caused by excessive neural discharges; in his own way, the doctor had been right. (Yet Galileo could not dislike him more.) The strong seizure the girl had suffered surely meant that most cells in the brain must have discharged together at high rates, quite unlike what happens normally, when only a few groups of nerve cells fire strongly, and the rest is rather quiet. So consciousness can fade not only if neurons are destroyed and cannot fire, as with Copernicus—consciousness also fades if all neurons fire at the same time. Somehow, if everybody shouts together, nobody can listen, thought Galileo.

"The girl would have been wasted anyhow," intervened the doctor, as if he could read Galileo's mind. "Nothing I could have done. She might as well serve a proper cause."

And himself, thought Galileo—could he have done something himself?

But then he thought: In this foul weather, in this foul land, how could one lend a helping hand?

NOTES

The discovery that changes in brain activity could be detected by observing changes in cerebral blood flow is due to Angelo Mosso, an Italian physiologist who, at the end of the nineteenth century, recorded the pulsation of the human cortex in patients with skull defects. As related by Marcus Raichle in "Historical and Physiological Foundations of Modern Neuroimaging" (*Psychological Science,* W. W. Norton, 2003), "Mosso came to study a peasant by the name

of Bertino who had incurred an injury that left him with a perma-
nent pulsating soft spot in his skull. One day while Mosso was using
some elaborate gadget to record these pulsations, the church bells
rang noon, and he noticed a sudden increase in pulsations over the
cortex. Mosso asked Bertino if he felt that he should have said his
mid-day prayers, and he said 'yes' and up the pulsations went and then
down again. So then, conducting what may have been the first cog-
nitive activation experiment ever, he asked him to multiply 8 by 12.
So Mosso poses the question and the pulsations go up and then down
again, and then Bertino answers and up and down they go again. And
from this he concluded that changes in circulation of the brain were
related to cognition." The use of camphor oil to induce seizures goes
back to Paracelsus. The painting by Cagnacci (*Maddalena Svenuta,*
slightly modified) is at the Galleria Nazionale d'Arte Antica, Rome.
Jeanne d'Arc and the peeping friar are from Dreyer's film *La passion de
Jeanne d'Arc.* Hers are also the eyes at the end, while the other eyes may
bear some resemblance to those of Josef Mengele, the Nazi doctor.
This chapter, which plays out in some northern country during one
of the many witch-hunts, contains references to a story told by Mann
in *Doctor Faustus* and to Kafka's *The Trial.* Galileo's final thoughts
strangely echo those of Friedrich Rückert, in *Kindertotenlieder:* "In
this weather, in this dread, / I should never have sent the children
out; / They have carried them away, / I dared not say anything about
it! . . . / Now these are just idle thoughts." *(In diesem Wetter, in diesem
Graus, / Nie hätt' ich gesendet die Kinder hinaus; / Man hat sie getragen
hinaus, / Ich durfte nichts dazu sagen! . . . / Das sind nun eitle Gedanken.)*

11

A BRAIN ASLEEP

*In which is shown that when cortical neurons
can be on and off only together, as during dreamless sleep,
consciousness fades*

How much experience is lost each time a dream is not recalled? How many lives were lived that cannot be remembered? And God, too, have we experienced and forgotten?

Galileo was meditating, when he felt light and heat come from a corner: on the far side of the room a stove was glowing, a military cloak and a dagger hung from the wall, and underneath a man was asleep, half-covered by a sack.

Frick had returned, and Galileo did as he was told: he shook the man's shoulder. Tell me what was on your mind just before I woke you up, he almost shouted. The young man slowly opened his eyes. "Nothing was on my mind," he said after a while in a drowsy voice with a strong French accent. "What is going on? I am half asleep." Was there anything on your mind? asked Galileo again. Did you see anything, hear anything, think of anything? "No," repeated the man. "I was fast asleep, and I wish you had left me there. And who are you to intrude on my sleep?" he mumbled, and turned to the other side.

Frick looked straight at Galileo. "Is it not remarkable?" he said. "You just witnessed a man emerge from unconsciousness, from utter nothingness." Why so remarkable? asked Galileo. It happens every night to all of us. It must be because the brain shuts down to rest. "That is the remarkable thing," said Frick. "While the man was sleeping, his brain was far from asleep. Thirty billion neurons in his cerebral cortex were busy firing just as intensely as when he was awake."

So what changed in his brain? asked Galileo.

"You see," said Frick, "when you are awake, the activity of neurons in the cerebral cortex is like a sea stirred by shallow, changing waves. But when you enter deep sleep, early in the night, it is like a sea swollen by deep, slow waves.

"Imagine now throwing a stone into the waking sea. The stone will trigger wavelets riding in various directions, like fast-moving ripples. But if you throw the stone into the sleeping sea, everything will be drowned by the deep slow waves, waves that carry with them, up and down, the entire sea.

"Similarly, when in the awake brain a neuron becomes active and begins to sing, its song is heard by its friends elsewhere in the brain. But when the brain is asleep, there cannot be any individual songs: as at the stadium, everything is drowned by the unison scream and silence of the crowd."

Is this then why consciousness fades? said Galileo, remembering the cerebellum: there consciousness could not bloom, because nobody could talk to anybody. In the sleeping brain, perhaps, consciousness vanished because everybody must behave the same way, so there was nothing worth talking about.

"Right," said Frick. "Think of it this way: the waking brain is like a pluralistic society—different groups of neurons have different allegiances and cast different votes. But when it falls into a dreamless sleep, the brain becomes totalitarian: everybody behaves like everybody else, they all fling their arms up and down together, and there can be no dissent. It is a monolithic brain, there is no freedom left, so it is no use to talk."

Maybe it was the same with the epileptic girl, thought Galileo. She had lost consciousness when all her neurons had begun to shout at the same time, like the crowd of uncouth villagers on the planks. Maybe it was like that: Galileo decided to ask the Frenchman again and went to awake him. But Frick restrained him: under half-open eyelids, the sleeper's eyes were moving fast from side to side, as if following the erratic flight of an insect. They watched him for some time. "Try now," said Frick, "you will be amazed." So Galileo shook the man awake again. And this time he was quick to raise himself and speak:

> "Oh slayer of wise dreams! My brain was hot with imagination. I was dreaming a most precious dream, and you murdered it before it could reveal its wisdom.
>
> "Let me recall, lest I forget everything: there were two books on a table in front of me. One was a dictionary, dry and useless, the other an anthology of poems. As it seemed to me, it contained at once the wisdom of poetry and the reason of philosophy. And when I opened it, I came upon a line from the ancient poet Ausonius: '*Quod vitae sectabor iter*—What path shall I take in life?' What path, indeed? And why?
>
> "A stranger appeared and said, '*Est et Non*—Yes and No.' Maybe that stranger was you . . . But then the intruder, the book, and the dream dissolved . . . and the question lay unanswered:
>
> "Am I to occupy my whole life in cultivating my reason—advancing myself in the knowledge of the truth, in accordance with the method I have prescribed to myself? Yes or no?"

"You see," said Frick to Galileo, "there are times during sleep when one is hardly conscious, like the first time you woke the Frenchman up, and other times when one can be just as conscious as when one is awake—he was that way in his dream. It all depends on how the brain is functioning. When his eyes were moving under the eyelids, as if inspecting the images in his dream, his brain was again a sea stirred by shallow, changing waves, much the way it was during wakefulness. Thus he was conscious, though unaware of the world around him. But when the waves become slow and deep, consciousness fades. And consciousness will expire forever once the winds of life die off and the water freezes."

The Frenchman had listened attentively, but now he demanded to know: Who were they, and what did they want? Why were they interested in his dreams? And how could they claim that when he was asleep the first time, he was not conscious? "I have never ceased to think," he said, "nor would I ever stop. God willing," he mumbled.

"But you said so," thundered Frick. "You told us there was nothing on your mind. Just like there will be nothing on your mind when you go to sleep forever."

The Frenchman was angry. "Consciousness never expires, not even for a moment. What the brain does is immaterial; consciousness uses it to communicate with the body, and through the body with the world, but consciousness is a different substance and does not need the brain to exist."

Then he composed himself and added calmly: "The brain is an extended thing, a thing that occupies space or moves in it. And it is made of parts. But consciousness is not an extended thing—it is a thinking thing. And it's not made of parts, but is a unity—you cannot conceive of the mind split in two as you can of material things: you cannot see the right side of what's in front of your eyes without seeing its left side, too, you cannot see the shapes of objects without seeing their color. So body and soul are two different substances, and there is no way that one can generate the other."

Then he continued: "I am a thing which thinks—a thing which doubts, understands, affirms, denies, wills, refuses, imagines, and feels. How could a mere mechanism ever give rise to thought and language, to doubt, reason, and will? A machine may be made to resemble us, take similar actions, even utter words, but it may never use speech in a way that indicates reason."

Frick looked scornfully at the Frenchman. "You will be surprised at what mere mechanisms can do, whether made of cogs inside a machine, or of thin flesh inside your head," he said. "It is easy to underestimate mechanisms—but life too is mechanism, though people did not believe it, until they were forced by overwhelming evidence. And mechanisms perform the most refined calculations, reason without a fault, play chess, and prove theorems. They can decide your course in life, whether you say yes or no, except you don't realize it. It was surely a mechanism in your head that made you a Christian."

The Frenchman did not reply, and Galileo interjected, as if speaking to himself: Can mechanisms explain my seeing the blue of the sky, or the glowing red inside that stove?

"Ah, there's the rub," said Frick. "Not thought. Not language. Not will. Those are the easy parts. The rub is in the spark—the spark of conscious light. Give me a spark of conscious light, and I will illuminate the soul." And he clenched his fist.

"That spark," said the Frenchman, "will flicker whether your brain is deep asleep or dead. Because that spark is immortal."

"Immortal! If you used your reason, you would know that you are dead wrong," exclaimed Frick, his eyelashes sticking out at the Frenchman. "The first time we woke you up, you said yourself that there was nothing on your mind. How could your consciousness have been there and yet you experienced nothing?"

"What a poor philosopher you are," laughed the Frenchman with undisguised contempt. "You think I do not know better than that? Of course consciousness is there all the time, during sleep and after death. It is just a certain kind of corporeal memory that disappears, so when you woke me up the first time, I could not remember anything. But I was there, my soul was there, and will be there forever."

"If you want to hide behind the claim that consciousness is always there and you just don't remember it, then hide," Frick said coldly. "There is no way to disprove that, and I am not going to waste my time with it." He exchanged a look with Galileo. Then he went on:

"I ask you this. Countless medical mishaps and clever experiments prove that consciousness depends, in all its aspects, in quantity and in kind, on the function of the brain. These are just facts, and you better pay attention to the facts. Do you accept them? Or do you renounce them? Thrice I shall ask you this."

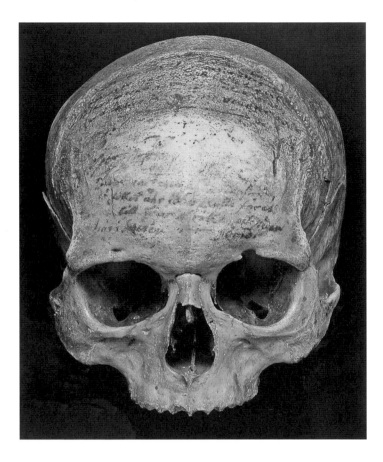

"What facts?" asked the Frenchman, with challenge in his voice. "I have studied facts all my life."

Frick was beaming. "I have studied my facts, too, and here is what facts say. You have a stroke in one part of your brain, and you lose consciousness of color. A stroke in another part, and you lose consciousness of faces. Consciousness does indeed have parts, just as the brain does—so much so that if I were to cut your brain in two, your consciousness would be split. And there is more. A stroke in yet another place, and you lose consciousness of the left side of the world: you see only the right side of the room, you eat from the right side of the plate, you dress your right side only, and you only shave the right side of your face. And if you have a stroke in some other region of the brain, you may be left without language, or reason, or emotion, without your moral sense, or without your will. You may even lose your belief in God, if the Devil places the stroke in the right spot."

The Frenchman stared at him speechless, and Frick went on. "Do you need more evidence, Saint Thomas? If an artery breaks inside your skull and the blood floods away most of your brain, you cease to exist. If you are poisoned with drugs, your self is annihilated. If your brain is taken by a seizure, you evaporate into nothingness. It is enough for the blood to briefly drain away when you stand up, and you faint unconscious. You enter sleep, as we just saw, and your consciousness fades away."

"I have no quarrels with the facts of medicine," said the Frenchman with an ill-formed smile.

"So tell me this then, and be absolutely clear," said Frick. "The facts say that, in all circumstances we know of, the working of the brain determines how conscious you are, and of what. But if your soul is immortal, what happens when your brain is dead? Will you be awake or asleep? Will you have the pleasure of sight? And in that case, what will you see? Will you have the privilege of sound, and what will you hear? Will you have desires or regrets, and what will you wish for? For if you had a soul without a brain, it would have to be empty."

"I will be clear," said the Frenchman. "I will be conscious of my soul's oneness with God."

"Why not of your soul's congress with the Devil, then, or of an intercourse with nothingness?" snarled Frick. "If you forsake logic, all alternatives are equally arbitrary." He glanced at the Frenchman again, and added scornfully: "And even so, what kind of oneness with God would it be, if it cannot be experienced, or remembered?"

There was a sudden silence. The Frenchman kept his lips tight, and so did Frick, but then he asked Galileo: "What do you make of these people? People who dream of God and want to fit reality to a dream, even if doing so means splitting reality in two, and murdering the facts, and renouncing reason?"

Galileo said nothing. But he thought that if you give in and sin against reason, even just once, everything is possible, and nothing.

NOTES

Sleep is probably the ideal test bed for studying consciousness: most people can testify that often, after a brusque awakening early in the night, one may emerge from virtual nothingness into the fullness of experience, just like Descartes in this chapter. This tells us not only that consciousness is not a given—it takes merely a change in brain activity and all is lost—but also that the difference between brain activity in wakefulness and in deep sleep (and perhaps some forms of anesthesia) may hold the key to how consciousness is generated. Unfortunately, the author gives a poor and crude account of the neurophysiology of sleep, presumably due to sheer ignorance. An experiment not unlike throwing a stone in the waters of the brain, first during wakefulness and then during sleep, is described in Massimini et al., "Breakdown of Cortical Effective Connectivity During Sleep," *Science* (2005). Except that the results of that experiment can also be interpreted as indicating that during deep sleep the brain splinters into disconnected modules, a bit like in the hysterical Teresa, just in a more extreme manner. The ship of consciousness in distress in the waves is slightly modified from a sea scene of Willem van de Velde (Worcester Art Museum, Massachusetts). The dreams and some quotes (modified, as usual) are all from Descartes. Rembrandt's *Philosopher in Meditation* is at the Louvre. Descartes's head is by Frans Hals (Louvre). Descartes's skull is courtesy of Muséum National d'Histoire Naturelle, Paris. Caravaggio's *Incredulity of Saint Thomas* is at Sanssouci, Potsdam. One would have a hard time deciding who is more arrogant between Frick and Descartes. Crick, it should be mentioned, was a devout atheist.

PART II

———

{THEORY}

Experiments of Thought

12

INTRODUCTION

The Enigma of Consciousness

A blank slate is the best map for tracing a new path—thoughts form more easily with nothing else to watch. So Galileo stared at a bare wall, rehearsing in his mind what he had seen, the rooms he had entered, the people visited. Frick was right: consciousness depends entirely on the brain. If the brain ceases to function, nothing is left: not us and not the world around. When Copernicus had lost his cerebrum, the universe had lost his point of view, and his own universe had vanished to a point.

Not all of the brain's vast continents matter, only a few privileged lands. The trembling painter had lost as many nerve cells as Copernicus, yet he was fully conscious. So the cerebrum matters, but not the cerebellum—how neurons are connected, not just how many of them. The cerebellum is like a great archipelago—millions of islands separated by an impassable sea. If every isolated island, each with its own village, were submerged by a cataclysmic sea wave, who would know or notice? But if Europe, the cerebrum of the world, had been wiped out, all would had been lost. Why was it so? Wasn't the cerebellum just as populated as the cerebrum?

He had learned, too, that the nerves that originate in the eye and reach the cerebrum are not essential for consciousness. The blind painter's eyesight was dead, but not his mindsight: in his blindness he could still imagine colors and shapes, and dream of them; he could project bright images on the walls of his darkened mind. Neither was there any need for the nerves that leave the cerebrum and steer our every deed: his friend M. lay mute and paralyzed, inert as an ancient rock, but was as conscious as his visitors were. Not even the intricate nerve loops that leave the cerebrum, perform complicated calculations, and bring back the results are necessary for consciousness, though without them consciousness may lose her faithful slaves—those that relay messages to their empress and interpret her every wish, and those that store and retrieve her memories—and he thought of the poetess and the gamba player.

Then he remembered Ishmael, and Teresa. They had taught him that consciousness can be divided if the connections linking the two halves of the cerebrum are split, or if the front is separated from the back, whether by the icy water that Salerno had poured, or by the hot vapors that rose from the womb.

He thought of the wretched girl on the wheel—what he had learned from her. When the doctor poured camphor oil onto her brain, he had learned that consciousness can be lost, as well, when too many nerve cells are active together. And from the Frenchman asleep near the stove, he had learned that when the waves that stir the ocean of the cortex are too deep and uniform, consciousness is drowned. So consciousness depends not just on certain parts of the cerebrum, but those parts must be functioning in the proper way.

But then, what is special about those parts of the cerebrum, and the way they function when we are awake or dreaming? How can it be that the sweeping universe of consciousness is due to a fistful of jelly inside a skull? And only when it quivers of quick, intertwining waves?

Galileo was searching for answers, but Frick was nowhere to be seen. Instead, a stranger entered the room. His name was Alturi, or so it sounded to Galileo.

He addressed Galileo so: "You've learned your facts about the brain and consciousness, or so I shall presume, but now tell me: How much wiser are you for all that? Remember, the world is full of facts and short of logic." Then, with no hesitation, Alturi said: "I know what's troubling you. Just speak the question, as simply and clearly as you can. One goes quite far with logic and imagination," he added, and seemed confident in the extreme.

For a while, Galileo sat staring motionless at the blank wall. But Galileo was not one to turn his back to a challenge. There were really two problems, Galileo had concluded. They could be called the first and second problem of consciousness. So he stood up, facing Alturi, and went on to state them:

Here was the first problem of consciousness: What determines if consciousness is present, and where it lives? This was the problem Galileo was struggling with right now. Why does consciousness live in the brain and not in the liver? Why, for that matter, only in certain parts of the brain—the cerebrum and not the cerebellum? The world contains many material things, galaxies and stars and planets, mountains rocks and sand, oceans lakes and rivers, oaks and wheat and daffodils, towers mills chairs and clocks, limbs and trunks, lungs and hearts, ears and eyes, but consciousness seems to live only inside a bright point in every person's head.

There was a second problem, too, one that should be tackled, perhaps, only after the first one was solved. Here was the second problem: What determines the specific way consciousness is? What determines the visual quality of sight, or the auditory quality of sound? There must be something about the organization of certain parts of the cerebrum that makes each of them contribute specific qualities to consciousness, said Galileo. But precisely what aspect of neural organization is responsible for shapes to look the way they do, and different from the way colors appear, or pain feels?

Alturi smiled faintly. "Trust me, there is really just one problem. If you knew what determines whether consciousness is present and where it lives—why there is one inside my head, one inside yours, and none in a heap of sand—you would not be far from knowing what determines its quality, why it feels this way or that way, why it feels purple or painful. One problem," said Alturi, "though hardly an easy one. In fact, so hard that it may have no solution, in practice or in principle. So hard as not only to resist science but to defy imagination—so hard that we cannot envision how it might possibly be solved."

He turned to Galileo and went on, as if to explain. "Maybe one day we'll know, to the last exacting detail, how the brain spins its wheels. But this will give no traction to climb the peak of consciousness. Take memory," continued Alturi. "To understand completely where memories are stored, and how, will be quite intricate, you can be sure of that. But it's conceivable—there is no lurking mystery, and it's being done. Indeed, machines can do it, too, machines that are the progeny of prototypes I had built myself.

"Similarly," said Alturi, "we may be in awe at how the brain controls the refined sequence of motions it needs to play an instrument. We may be in awe at how it strings together sounds to make its words and words to make its sentences. But I can prove to you, it's merely a matter of logic: every such feat can be reduced to a series of simple steps, and if it can thus be reduced, then a machine can execute it.

"Take the action by which you grasp an apple and bring it to your mouth to take a bite, whatever your reasons for doing so," said Alturi: "The programming of the motor sequence, the perfect coordination among the various limbs, the extraordinary control of forces brought into play by dozens of different muscles, the compensatory adjustments of posture—these are remarkably complicated feats. However, they are fully within the reach of a machine—nothing is mysterious or unfathomable. The brain, this enchanted loom of a billion billion wires, can certainly house such a machine. Once we have figured out the mechanisms, there will be nothing left to explain.

"Likewise," continued Alturi, "we may wonder how the brain can intercept mere flutters in the air and extract words and sentences out of them. Think how a foreign language sounds before you understand it—one long stream of clatter, with no beginning and no end. Yet the brain soon learns to parse words, one neatly separated from the other, and attaches a meaning to each of them. Or remember what you were told: the brain can recognize an object even though its shape may change depending on our point of view. It can recognize the color of an object even though the light in the room might be the warm light of sunset or the cold light of lightning. These are all remarkable feats, but there is nothing mysterious about them. Indeed, a machine can be built to do just as well as we do, a machine that can recognize objects, identify each color, recognize red or blue every time we do. It is merely a matter of having the machine go through the right series of computations. Certainly, the visual parts of our brain must contain such a machine. How else would the brain do it?

"But then there's the rub," said Alturi, as Frick had said, and paused for a while. "You see," said Alturi, "even if we understood completely the mechanisms for distinguishing among the colors, down to the most niggling detail, we would still fail to understand why, alongside the nerve cells in the visual parts of the brain, which slavishly perform their task—the task of distinguishing among colors—and use those distinctions to guide our behavior, why alongside all of this there should also exist a conscious experience of color—an I who sees in front of himself, vividly, a red apple, or a blue sky. No matter how I program the machine, what steps I force it to follow, I cannot see how it could see the same way I see," said Alturi. "It could perhaps behave like me, but I just cannot see how it would experience anything at all." And he smiled again, faintly.

Then he showed Galileo a letter he had received from a famous brain scientist:

When I turn my gaze skyward I see the flattened dome of the sky and the sun's brilliant disc and a hundred other visible things underneath it. What are the steps which bring this about? A pencil of light from the sun enters the eye and is focused there on the retina. It gives rise to a change, which in turn travels to the nerve layer at the top of the brain. The whole chain of these events, from the sun to the top of my brain, is physical. Each step is an electrical reaction. But now there succeeds a change wholly unlike any that led up to it, and wholly inexplicable by us. A visual scene presents itself to the mind: I see the dome of the sky and the sun in it, and a hundred other visual things besides. In fact, I perceive a picture of the world around me.

"This is the enigma of consciousness," said Alturi.

NOTES

The picture is of Alan Turing, who formalized the notion of computation by conceiving the Turing machine (a thought experiment) and developed the idea of the universal computer, which he then helped to build. During World War II, Turing worked at breaking the German codes and was able to unscramble that of the powerful Enigma machine. He was convicted of gross indecency after admitting to a sexual relationship with a man and was required to undergo hormone therapy. Two years later he committed suicide by eating an apple laced with cyanide. The quote at the end is from Charles Sherrington, a neurophysiologist who discussed consciousness in *Man on His Nature* (Cambridge University Press, 1951) and compared the brain to an enchanted loom. (The loom is from an unidentified source.)

13

Galileo and the Photodiode

In which is shown that the humble photodiode
can tell light from dark as well as Galileo

"Do you see a way out of this dilemma? One that does not substitute an enigma for a mystery?" asked Alturi, leading Galileo through a side door. Alturi wanted him to perform a thought experiment. After all, Galileo was the champion of thought experiments: Was he not the first to make use of them in the service of the new science? Galileo remembered word by word how he had managed to establish the relativity of motion:

> *Shut yourself up with some friend in the main cabin belowdecks on some large ship, and have with you there some flies, butterflies, and other small flying animals. Have a large bowl of water with some fish in it; hang up a bottle that empties drop by drop into a wide vessel beneath it. With the ship standing still, observe carefully how the little animals fly with equal speed to all sides of the cabin. The fish swim indifferently in all directions; the drops fall into the vessel beneath; and, in throwing something to your friend, you need throw it no more strongly in one direction than another, the distances being equal; jumping with your feet together, you pass equal spaces in every direction. When you have observed all these things carefully (though there is no doubt that when the ship is standing still everything must happen in this way), have the ship proceed with any speed you like, so long as the motion is uniform and not fluctuating this way and that. You will discover not the least change in all the effects named, nor could you tell from any of them whether the ship was moving or standing still.*

And then he remembered how he had undermined the very foundations of Aristotle's physics and formulated the first law of dynamics—the principle of inertia—relying not on conclusive experimental proofs (which would have been difficult to acquire) but simply on the power of his imagination. So he had imagined a perfectly spherical ball, moving on a perfectly smooth plane, never encountering even the slightest resistance. And by imagining this ideal situation, he had concluded that the ball could do nothing other than continue to move uniformly, requiring no further force to maintain its motion.

So what was Alturi thinking of? A thought experiment to reveal the essence of consciousness?

The side room was exceedingly bright, unfurnished except for a comfortable chair in the middle. Alturi invited Galileo to put himself at ease. Sitting on the chair, Galileo at first felt dizzy, as there was nothing upon which to focus his eyes. In front of him were no discernible walls: it was as if a boundless white screen encompassed his field of vision. The dizziness went away, and all that was left was the perception of pure light.

Alturi quickly explained the experiment. Galileo was simply to report whether there was just light, as it was now, or whether everything plunged into complete darkness. Galileo need only say "light" or "dark," depending on the circumstances—nothing more was required. Alturi urged him not to think, not to allow his mind to wander; and instead to concentrate on which response he should provide—light or dark—however simple that task might seem. And then Alturi left.

Indeed, soon thereafter everything became dark, but for an instant only; then it turned light again, then dark, then light again. Galileo had to respond rapidly, as light and dark alternated as fast as he could speak, and report "light" or "dark." Galileo was diligent and announced each transition.

After a few moments, Alturi was back in the room, inquiring whether Galileo had seen both light and dark. The task was so simpleminded that it took no brains, thought Galileo. Did Alturi

mean to imply that Galileo was blind? Was Alturi aware of his
failing eyesight? Had Galileo signaled incorrectly a few times? Of
course he had seen light and dark, dark and light. What could have
been easier? What was the significance of the experiment, if there was
any?

Alturi said nothing but pointed to a thin line of wire attached under
the chair. The wire came from a small hole in the floor, and at its tip,
facing toward the front of the room, it carried a little glass bulb. That
was the photodiode, Alturi said, one of the simplest machines one
could possibly build—a supremely simple object. "A photodiode is
nothing more than a miniature electrical circuit containing a variable
resistance," he said. "What causes the resistance to vary is exposure
to light, so that the flow of current increases when the light becomes
more intense."

And so he explained that, all along, while Galileo was shouting
whether he saw light or dark, the photodiode too had been doing
the same: whenever the light had turned on, making the current rise
inside its only circuit, the photodiode too had sent an "on" signal and
an "off" signal whenever the light had turned off, making the current
drop to zero.

From outside the room, Alturi had been recording both Galileo's
responses and those of the photodiode. As far as he could tell, he said,
the responses of the photodiode were every bit as accurate as those of

Galileo—indeed, their responses were indistinguishable—they had both passed the test.

The purpose of this strange duel between the great Galileo and a mere photodiode was now dawning on one of the contenders. It seemed to Galileo that the question he must ponder was simple enough. Was the photodiode conscious of light and dark, the way Galileo himself was? Did the photodiode have the privilege of subjectivity, did it *experience* the qualities of light and dark?

Galileo did not even stop to think: the answer must be no.

"How can you be so sure?" interjected Alturi at once, as if savoring an impending victory. The experiment proved that the photodiode did just as well as Galileo at distinguishing between light and dark, if not better. On what grounds, then, could Galileo conclude that the photodiode would not *see* the light and the dark just as well?

Galileo knew that the photodiode was nothing more than an electrical circuit that could only change its state—on or off—according to the lighting. That could never be enough to engender a conscious experience of light and dark, of this he was certain. But why not? What was the critical difference between the photodiode and himself, if there was one?

After all, he had been reading eagerly every single book Frick had given him, and he had learned that the nerve cells in his own brain were not that different from photodiodes. He had read, for example, that the photoreceptors in the retina behave just like sensitive photodiodes, though they are made of flesh rather than metal and though the mechanisms by which they respond to light are different. True, these receptors were connected to other cells, and these in turn to other cells, until they reached the highest levels of the brain. But he remembered Frick telling him that cells higher up in the brain, no matter how high up, were still behaving very much like photodiodes. And indeed our capacity to experience light and dark depended on those photodiode-like cells, and we would lose it if those cells were destroyed.

Galileo now saw the thrust of Alturi's argument. His own capacity to distinguish between light and dark was based on certain cells in a certain part of his brain. In principle, these cells were little more than photodiodes made of flesh. Given that a human being seems to have a conscious experience of light and dark depending on the activity of photodiode-like cells somewhere in his brain, then why shouldn't the photodiode, too, have a conscious experience of light and dark?

Galileo was confused, but something made him hold firmly on to his belief: a photodiode cannot be conscious of light and dark the way a human being is. He could not come up with a satisfactory explanation, Galileo knew that, but an explanation must exist, of this he was quite sure.

As he was sure of one of the foundations of his thinking, perhaps the most important one: the principle of sufficient reason, *Nihil est sine ratione cur potius sit quam non sit*—there must be a reason why things are the way they are, and not another. This was a principle that Galileo had always tried to follow. There must be, thought Galileo, some necessary and sufficient conditions whereby the activity of certain cells located in the visual cortex is associated with a conscious experience of light and dark, whereas the activity of the photodiode is not. But which conditions?

This, then, perhaps in its simplest form, was the first problem of consciousness.

NOTES

Turing, a pioneer in the debate on artificial intelligence, proposed the Turing test to find out whether a machine can think and perhaps be conscious. He suggested that an intelligent machine and a human be set up in one room, and that they should engage in "teletyped" conversations with a human interlocutor in another room. If, after a given time, the human interlocutor could not tell which of the conversation partners was the machine and which was human, then the machine had passed the test and could be called intelligent. Here Alturi obviously meant to administer a Turing test to Galileo, albeit the simplest possible one. The quote is from the *Dialogo sopra i due massimi sistemi del mondo tolemaico e copernicano* (Florence, 1632). The Chinese man in the room with file cabinets is Emperor Yongzheng, whose chest betrays Charles Babbage's analytical engine, one of the first mechanical computers. The philosopher John Searle laid out the Chinese Room argument in "Minds, Brains and Programs" (*Behavioral and Brain Sciences,* 1980). There he argues that a machine following a set of instructions would not think, or be conscious, or enjoy meaning the way we do, even if it were able to answer all kinds of questions appropriately and thus pass the Turing test. He shows this by imagining that inside the machine there is actually John Searle, who carries out the instructions without understanding a word of Chinese: "You're locked in a room with a bunch of Chinese symbols on cards and you have a program which tells you how to give them back through a slot in the wall in response to other cards coming in, and all the same you don't understand Chinese." The argument is used by Searle to make the point that one cannot get meaning (semantics) out of mechanism (syntax), but it can also be used to claim that one cannot get consciousness out of computation. Indeed, Galileo will come to the conclusion that there is no meaning without consciousness and that, in a fundamental sense, consciousness and meaning are one and the same thing.

14

INFORMATION: THE MANIFOLD REPERTOIRE

*In which is shown that the repertoire of possible experiences
is as large as one can imagine*

What was the essential difference between the photodiode and himself? thought Galileo.

Why did he see light or dark, why was he conscious of it, whereas the photodiode just responded to it, like the simple machine it was? What was the critical difference?

And suddenly the space became bright again, but this time it was not blank: Galileo saw he was immersed in a uniform, intense blue light. All was blue.

And then, just as suddenly, all turned red. Then green, and yellow, and soon he was seeing the space brightly colored, of one shade of color after the other, thousands of them, and then it was colored of all those shades at the same time.

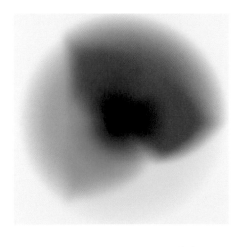

And then something remarkable happened: in front of his eyes, what was before a featureless space became a wall, and then a painting appeared on the wall. It was a portrait, one he recognized all too fast—how young she was then—but before he could feel his old heart rest again in the grave of his chest, she had transformed into another face, and then into another—one face after the other, at increasing speed. At first he recognized them, but then faces appeared that he had never met, and some wore strange hats and even stranger hairdos.

He might have seen a thousand faces, and then the painting changed and seemed to transform into a window. And the window showed a familiar old scene. It was Pisa, where he had studied medicine, and hated every part of it. But then suddenly it was not Pisa but Padua, where his discoveries had brought him fame, and then Rome, where his downfall had begun. And then it was countless other cities, and villages, and places, and palaces, and other rooms, and gardens, and mountains and valleys, most of which he had never seen.

And then through the window the scenes were changing at even greater speed, and soon he was hearing voices outside, known and unknown, in languages known and unknown, saying something different every time, or even the same, but in a different tone, faster and faster, and then it was sounds, and chords, all possible chords, and noises he had never heard.

And then he started smelling the herbs of his garden, one after the other, sweet or pungent, and the smell of oil, and that of old books, and that of death, and then, though he was not drinking, he tasted his wine, and then other wines, he could not tell how many different ones, ones that were sizzling, and dense syrups, and spices, too, but soon it was fruits, fruits he knew and many more he did not, and then chestnuts, and almonds, and there was more, there were dishes old and new, and flavors of dishes he did not know existed, and of game he could not imagine, strong and unique.

And then he felt fear—fear that he was losing his mind—and he felt angry and then grateful, and then at peace, and then he felt a loss so acute it pierced him like a knife in the heart—as when he had learned she was no more—and then the thoughts started, and they were too many, his mind was wandering from one to another, and he recognized his own thoughts, and then thoughts he had never had, he thought of his own grave and thought it would be in Florence and then that it would not, and then he saw it, and saw his own body, and thought that he was separated from it, and so fast was he thinking that he became dizzy of thought, and the thoughts were revolving in his head, and they were going at such speed that he could not follow them, and then he felt empty and then collapsed.

And then—it could have been an instant, or it could have been forever—he saw Alturi: he had entered the room again and had a smile on his face like an enigma.

"You see what this means," said Alturi. "When a man tells you he loves you above all others, you think you are special, and he is an angel. But if you happen to be the only man in the village, you are not very special at all, and he does not mean much by that."

Galileo did not understand.

"You remember when the light turned blue? What did you think the photodiode was signaling when you saw blue?"

It could not possibly signal blue—all it could signal was light or dark, said Galileo.

"Precisely," commented Alturi. "And after that, what did you experience?"

Far too many things for me to remember, answered Galileo. It all went red, then green, then I went through colors, and shapes and faces, and scenes and places and sounds and thoughts. My mind went spinning, and I confess, it has not yet recovered.

"And what do you think the photodiode did?" asked Alturi again.

What could it do? Whenever the image had enough light, it would have signaled light, and whenever it did not, it would have signaled dark. And when there were sounds, and smells, and pains, it would have gone on saying dark. Would it not?

"Of course it would. Precisely," said Alturi, his face without expression.

Just then a man entered the room riding a strange bicycle with a single big wheel. The man was juggling balls and smiling absentmindedly. His name was S.

"Gentlemen, let us not waste our time with images," said S. "Let's not beat around the bush. What you need is simple, you need a formula: p log p, gentlemen, a formula for information!"

S. came off the unicycle, still juggling. "That will tell you how large is the repertoire of possible states. Take the photodiode, gentlemen. What's the uncertainty about which state it actually was in, if you do not know anything at all? Think about it. Its repertoire of possible states is small: either it was on or it was off. Now say you obtain some information—that's the word, gentlemen, informa-

tion! Say you learn the photodiode was actually off. No matter by which means or mechanisms this information was generated, it eliminates the uncertainty: now you know the photodiode was off rather than on, no uncertainty left. That's what information is, gentlemen, reduction of uncertainty. And with the photodiode, where complete certainty means ruling out just one state out of two, the amount of information is just one tiny bit.

"On the other hand, gentlemen," he said quickly, turning around and letting the balls fall by his side, "if the repertoire of possible states is large, say as large as that of your brain, then there is great uncertainty about which state it was actually in. So now again, if by some mechanism it is established that your brain was in a particular state, and not in a trillion others, that's lots of uncertainty eliminated. That's many, many bits of information, gentlemen—bits, I say, because that's how information is measured by my formula, p log p— beware of imitations."

S. mounted his unicycle. "Just so you know, I also have the formula for juggling. Oh, and I almost forgot," he giggled, "information is a number, so do not ask it for meaning, never mind consciousness." He swerved and rode away.

Perhaps, thought Galileo, but what did S. mean, and what use was his formula? And then it dawned on Galileo. Perhaps the essential difference between the photodiode and himself was this—every time he, Galileo, had a vivid experience, even the simplest one, such as pure darkness, his brain was not merely distinguishing one possibility from another, his brain was not just telling dark from light (though Alturi had set him up like that). No, his brain and its complicated mechanisms were distinguishing between pure darkness and countless other situations, which would have led to trillions of different experiences. Because to Galileo, dark was not just different from light, it also meant it was not red, or blue, or any color in the rainbow, any face, any place, any sound and smell and flavor, any feeling and any thought, and any of their combinations.

But to the photodiode, dark must have meant much less. With its simple mechanism, the photodiode had no way of knowing that dark was not a color, not a face and not a place, not a sound or smell or flavor, not a feeling and not a thought—to a photodiode, dark was not dark, but merely one out of two. To a photodiode, the whole universe merely was: *this or not this*. Maybe the essential difference between the photodiode and himself, indeed, was information.

So Galileo came to a simple thought: perhaps it was due to the immense repertoire of alternatives that his brain could distinguish that he was conscious and the photodiode not, or infinitely less so.

Consciousness makes up in number for what it lacks in weight, he thought, remembering Sanctorius. This realization was so obvious that Galileo wondered why it had not occurred to him before.

NOTES

The painting is Titian's *Portrait of a Woman* (Louvre). The encased photographs are of the Battistero in Pisa, of Sant'Antonio in Padua, of Villa Medici in Rome, and of Galileo's tomb in Santa Croce, Florence. The still pictures are from Godfrey Reggio's *Koyaanisqatsi*.

Claude Shannon, an electrical engineer and mathematician, is the father of information theory. In his work (and in the chapter), Shannon intimates that information is a number and is divorced from meaning. Just as science flourished once Galileo removed the observer from nature, communication and storage of data exploded once Shannon removed meaning from information. Later on Galileo will begin to think that Shannon's prescription may be correct only from the "extrinsic" perspective of an observer. However, information integrated by the causal powers of a mechanism inside a system, from the system's "intrinsic" perspective, acquires meaning—in fact, it becomes meaning. And the observer is returned to nature. . . .

This is Shannon's p log p formula, where S stands for entropy:

$$S(X) = -\sum_{m=1}^{M} p_m \log_2 p_m$$

X can be any system that can take one of a number of states $m = 1 \ldots M$: in the case of the photodiode, the states are 1 to 2. Every state has a probability p_m (½ for the photodiode's states), and the sum of all probabilities is 1. If all states are equally likely, the formula reduces to the logarithm of the number of states (also known as Boltzmann's formula). If some states are more likely than others, the formula takes into account the distribution of probabilities of each outcome, and the uncertainty is reduced. Formulas based on Shannon's entropy, such as mutual information and relative entropy, the distance between two probability distributions, can be used to measure information as the reduction of uncertainty. Shannon's magnetic mouse Theseus was the first device (controlled by relay circuits) that was able to find his target in the maze by learning through experience (courtesy Alcatel-Lucent USA Inc.).

Aside from his work on information theory and communication, Shannon did work on an equation for juggling. He also built unicycles and loved to ride them through the hallways of Bell Labs. (The unicycle photograph is from the Library of Congress.) The Most Beautiful Machine is also an idea of Shannon's. When the "on" button is pushed, the trunk opens, a hand comes out, turns the machine off, and the trunk closes (from *Rolling Ball Sculptures and Kinetic Objects* by Hanns-Martin Wagner).

15

GALILEO AND THE CAMERA

*In which is shown that the sensor of a digital camera
has a large repertoire of possible states,
perhaps larger than Galileo's*

To make progress, one may need to step back, thought Galileo, look over one's shoulder, and not take things for granted.

Consciousness we take for granted, he thought, because we always had it, and it requires no effort. We see dark, we see light, we see a woman, we see any of a trillion things—they are just there, she is just there, immediately there, with no need for us to seek, compare, or calculate. And yet that immediacy may be illusory, and yet she may appear to us only because we are inordinately rich, because our brain can pick and choose from an inexhaustible repertoire—the repertoire of a thousand lifetimes. If we did not, if we had the insignificant repertoire of a photodiode, maybe we would not see her, we would not even see the dark—perhaps we would see nothing at all.

But Galileo's reflections had a short life: his hands were trembling, not of an old man's weariness, but of a child's thrill—they were holding the camera as they once held his first telescope. Alturi had given him the digital camera, and he had learned to capture digitized images like an instant painter, to send them over waves in the air, to store millions of them on a pinhead, and to show each painting on a large, beautiful frame where it was created anew. He could take pictures of anything—from a galaxy to a grain of sand. He could capture the effigy of all and everything—pictures of everything, of Pisa and Padua and Rome, of faces he knew and others he did not, and even of himself.

He had not imagined any of this. How could he? He had not imagined one would go so far so fast—that men a few generations younger than himself would know so much—their powers so astounding.

There is no imagining where science can lead, thought Galileo. And yet he had imagined that it would lead far away, and he had taken the first steps. They were small, but they were bold, he said to himself, and felt like a proud father who had not seen his child and, decades later, had found her a magnificent queen.

But Alturi was there to question, not to let him play. He had explained to Galileo how the camera worked. Its heart was a sensor that contained a million photodiodes, arranged in a square grid just behind the lens. Like any other photodiode, the photodiodes on the camera sensor signaled light by an increase in current, each of them for a different element of the image.

This was Alturi's question: Was the camera conscious?

Alturi clearly had a point. To be conscious, Galileo had concluded, a system must be able to distinguish among a large repertoire of possible states. Then a photodiode, with a repertoire vanishingly small—just one state corresponding to dark, and one corresponding to light—could only be minimally conscious, indeed just one bit conscious.

But Galileo had not considered what an array of one million photodiodes could do. Because it was clear that one million photodiodes—a camera—could do as well as Galileo: distinguish among as many images as he could, and perhaps more. And if that was not enough, one could build a larger camera, and soon it would beat Galileo at his own game.

Galileo knew exactly where Alturi was leading him. If each photodiode could distinguish between just two possible states, corresponding to light or dark, an array of one million photodiodes could distinguish among $2^{1000000}$ possible states. This was a number so large that it dwarfed the number of stars in the sky or that of grains of sand in the sea: the repertoire of the camera sensor would be worth one million bits.

"So is the camera conscious?" asked Alturi.

No, said Galileo, showing no desire to say more.

"But its repertoire of possible states, or shall we say of possible images in this case, is at least as large as yours," insisted Alturi.

Nobody has ever counted the number of possible experiences that are available to me, said Galileo.

"Then you should examine this," responded Alturi, and showed him a flickering screen of white and black dots. "What do you see?"

I see just a flickering mixture of white and black dots, like salt mixed with pepper, said Galileo.

"In reality you are seeing thousands of different pictures—some dots that are white in one picture turn black in the next, and so on," said Alturi. "But your brain cannot distinguish among them, and thus they all look the same to you. Yet to the camera, or to your retina, they are all different—each one of them produces a different image on the sensor. So the repertoire of the camera may well be larger than that of your own consciousness."

So it was, and so his thinking had been trivial, mused Galileo within himself. What he had concluded—that a large repertoire of distinguishable states is necessary for consciousness, that it made all the difference between himself and the photodiode—was a trivial fact: it was not even enough to make the difference between himself and the camera. So the size of the repertoire could not be the source of experience; and so information, too, as S. had said, juggling from his unicycle, had nothing to do with consciousness.

Galileo tried to focus, but his mind was wandering. An aging man he was, toiling with new confusing instruments, but facing the same old problems. He had walked into blind alleys before and knew when one should forgo the quest: better to lose faith early than late.

"Lose faith? What shameful ignominy for the great Galileo," screamed a short, crooked man appearing from nowhere. "First challenged by a photodiode, now fearful of a mere camera," said the man, whose name was K. "Fooled by his own perspective, Galileo himself believes the image on the camera is one. He thinks it's one! He does not see, the visionary physicist, the man armed with the telescope, that it is one only in the eye of the beholder."

K. went on: "Here I am, facing the scientist who began the mathematical study of nature, who removed the mind of the observer from the description of the world of physics. And I wonder, how could such a scientist miss the obvious, the self-evident, that the observer must be removed once more? And I marvel, rather, I lament," insisted K., "why is it always the same mistake, an error not of category, but worse still, an error of perspective? Yes, an error of perspective! Everyone makes the same mistake, people who should know better. I mean, they seem not to understand! That epistemologically, what am I saying, ontologically, even deontologically, one might add, the only true perspective, the one and true perspective, is the intrinsic perspective, is it not?

"Galileo," he then said, "the mind who first envisioned the relativity of motion, how could such a mind not see it? If one removes the observer—the observer who makes a unity of what is not—if one does that, the camera as an entity disappears, breaks down into its million constituting elements. For what exists truly, in and of itself, *per se* and *in se,* is not one entity with $2^{1000000}$ possible states, but one million separate entities, each with just two states. Such a simple truth! And yet even philosophers miss it. They seem to lack, so to speak, synthetic power! It even shows in their countenance!" And K. pointed to some portraits on the wall.

Prompted by K., Galileo examined the portraits with a magnifying lens. Indeed, under the lens, each picture soon dissolved, each face disintegrated. A million little dots appeared—each dot was what a photodiode knew of its subject.

"So now you see," said K. "But not to worry, there are better philosophers than those. I'll put your memory and mine to the test: Who said the following?

"That being is . . .
Indeed it is the same to think and to be . . .
Now that all has been named light and dark . . .
Everything is full at once of light and of dark night . . .
The One, if it has being, is One and Many."

Having recited those lines, K. showed Galileo a detail from a fresco by Raphael in Rome. It was the face of the only opponent whom Socrates had not trounced—the only one in all of Plato's dialogues.

Parmenides, exclaimed Galileo.

"Quite right," said K. "And what about this one?"

I know this one, too, said Galileo. It is Cardanus, who wrote a treatise on dice—perhaps the first that dealt with probabilities. (Cardanus, who, he thought, had also been condemned a heretic, had been incarcerated, and finally had come to abjure his theories.) But Cardanus had written another treatise, recalled Galileo, called *De Uno*—Of the One. There Cardanus had said that what makes a man a man is not his parts but rather their union.

"One more," said K., and Galileo saw a portrait of the Frenchman he had surprised asleep. He, too, Galileo recollected, had said that consciousness is a unity, not made of parts, that you cannot split it in two as you can material things, and that's why body and soul are two different substances.

"Excellent," said K. "Though I think I have said it best myself." And he handed Galileo a hefty book, open on a page that began like this:

The synthetic unity of consciousness is, therefore, an objective condition of all knowledge. It is not merely a condition that I myself require in knowing an object, but is a condition under which every intuition must stand in order to become an object for me. For otherwise, in the absence of this synthesis, the manifold would not be united in one consciousness. Although this proposition makes synthetic unity a condition of all thought, it is, as already stated, itself analytic . . .

What did K. mean? Galileo read further:

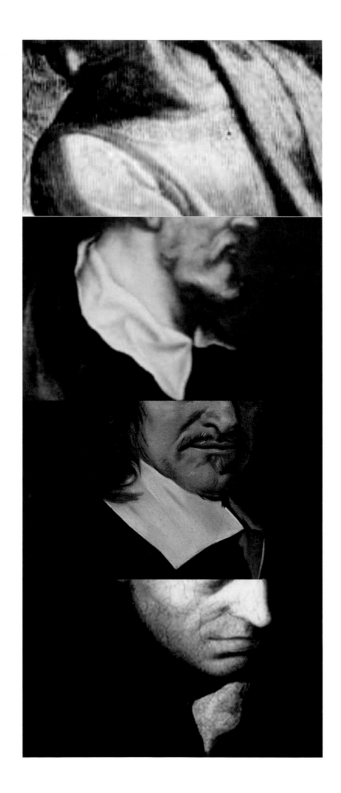

*The transcendental unity of apperception is that unity through which all
the manifold given in an intuition is united in a concept of the object . . .*

The transcendental unity of apperception? What was K. trying to
say? Why was it so hard to fathom? Galileo turned, ready to ask, but
K. was gone.

Galileo turned to Alturi, who had been listening and seemed
bemused. K. might be right, said Galileo, but why should it be said so
obscurely? Murky thoughts, like murky waters, can serve two pur-
poses only: to hide what lies beneath, which is our ignorance, or to
make the shallow seem deep.

For once, Alturi nodded. "K. is bad enough, but by no means the
worst of them," he laughed, and then he added: "Avoid philosophers.
In the fog of thought, banality dresses like a mysterious mistress."

Perhaps, said Galileo, but so does truth itself.

NOTES

A long tradition, atomism, holds that everything can be explained by
simple elements and their interactions. Is a picture just a collection of
points? As Galileo realizes, that is exactly how a camera sees a picture,
as a collection of points, but then, of course, it does not really see it.
On the other hand, each photodiode in the camera sensor should see
its point independently of the others; otherwise the camera would
lose information about the picture.

The four atomized portraits of atomists are of Democritus
(Velázquez, Musée des Beaux-Arts, Rouen), of Lucretius (unidentified), of La Mettrie (by J. C. G. Fritzsch, private collection), and of
Descartes, who was an extreme atomist/reductionist with respect
to the body (Frans Hals, Louvre). The four pictures of holists are of
Parmenides (from Raphael's *School of Athens*, Vatican Museums in
Rome), of Cardanus (unidentified), of Descartes, who was an extreme
holist with respect to the soul, and of Kant (unidentified). Among
contemporaries, Semir Zeki thinks that individual pieces of the cerebral cortex generate individual pieces of consciousness—their own

microconsciousness. The next quote is a composite from Parmenides in Diels-Krantz fragments II, VII, and IX, and Plato's dialogue *Parmenides*. The last one is from Kant's *Critique of Pure Reason*.

Hyeronimus Cardanus was a sixteenth-century physician, philosopher, mathematician, and avid gambler who made the first attempt to develop probability theory. (His was also the first study on dice rolling.) Cardanus was jailed for heresy (he had cast the horoscope of Jesus Christ) and managed to be released by recanting, which is probably why Galileo remembers him well. During the Renaissance, Cardanus, with Telesio, Patrizi, Bruno, and later Campanella, was among the first to put forth a panpsychist view of the universe (or, for Bruno, of infinite universes). Cardanus and Bruno especially emphasized unity as a property essential to mind or consciousness, and Bruno associated unity with "monad."

The Head of an Old Man, who bears a strong resemblance to Galileo, is by Jacob Jordaens (Trustees of the Weston Park Collection, U.K.).

16

INTEGRATED INFORMATION: THE MANY AND THE ONE

*In which is shown that consciousness lives
where information is integrated
by a single entity above and beyond its parts*

When is an entity one entity? How can multiple elements be a single thing? A question simple enough—but one, thought Galileo, that had not yet been answered. Or perhaps, it had not been asked.

The sensor of the digital camera certainly had a large repertoire of states—it could take any possible picture. But was it a single entity? You use the camera as a single entity, you grasp it with your hands as one. You watch the photograph as a single entity. But that is within your own consciousness. If it were not for you, the observer, would it still be a single entity? And what exactly would that mean?

While musing such matters, Galileo was startled by a voice. J., a man with the forehead of an ancient god, addressed him in a polished tone: "Take a sentence of a dozen words, and take twelve men, and tell to each one word. Then stand the men in a row or jam them in a bunch, and let each think of his word as intently as he will; nowhere will there be a consciousness of the whole sentence. Or take a word of a dozen letters, and let each man think of his letter as intently as he will; nowhere will there be a consciousness of the whole word," J. said.

Or take a picture of one million dots, and take one million photodiodes, and show each photodiode its own dot. Then stand the photodiodes well ordered on a square array, and let each tell light from dark for its own dot, as precisely as it will; nowhere will there be a consciousness of the whole picture, said Galileo.

"So you see that, Galileo," J. continued. "There is no such thing as the spirit of the age, the sentiment of the people, or public opinion. The private minds do not agglomerate into a higher compound mind. They say the whole is more than the sum of its parts; they say, but how can it be so?"

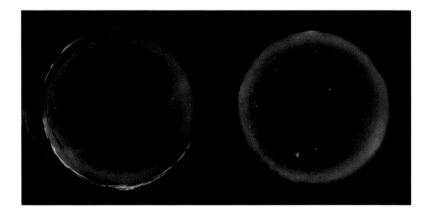

An image came to Galileo. An astronomer was watching the sky in Padua, during an eclipse, and precisely at the same moment, another astronomer was watching the night sky at the antipodes. Would there be a single consciousness contemplating, in one great image, the entire dome of the sky, the austral and boreal skies joined seamlessly at the horizon? A single image of the entire sky, experienced within one consciousness? That was absurd, thought Galileo, and its absurdity had nothing to do with the distance between the scientists. Whether the two were separated by the diameter of the earth, or by a fraction of an inch, like two photodiodes on the camera sensor, made no difference. Because in both cases, the two parties could not interact. And if they could not interact, they could not form a single entity, and they could not have a single, unified conscious experience.

"Of course," agreed J. "A blind man and a deaf man cannot compare sounds and colors. One hears them and the other one sees them, but could they compare them if they are together? Not even if they were to live in the same house forever, not if they were conjoined twins." Like Ishma and El, thought Galileo.

"Nice words," said Alturi, standing next to Galileo. "But what's the point? We were arguing about the information in a camera, and you saw that, if the camera is large enough, it can be as much or more than the information generated by a brain. Which shows that information, as S. was saying, has little to do with consciousness. Isn't it so?"

Galileo hesitated. If one measured information the way S. did, a camera was better than a brain: the larger the repertoire of states available to a system, the greater the reduction of uncertainty—the greater the information generated by the particular state the system was in. But was this the right way of measuring information? He thought of what J. had said, of the scientists in the northern and southern hemispheres, of Ishmael's left and right hemispheres. So he tried:

It should make a difference if the information is generated by a system that is one, rather than just a collection of parts.

"Quite possibly," said Alturi. "And how would one show the difference?" He smiled, as if he knew that Galileo could not provide an
answer.

"I wish I knew," said J., as if he knew there could not be an answer.

Galileo paused, as if lacking for words, then turned to J., and asked:

If, with an extraordinarily thin and sharp blade, old Occam's razor,
say, one were to split in two the sensor of the camera, in such a way
that half a million photodiodes lie on one side, and the other half a
million on the other side, what would then happen to the image seen
by the camera?

"Nothing would happen, of course," answered J. "The camera
would go on working just as well, taking full pictures, the pictures
could be sent over the air, stored and replayed at will, and no one
would notice any difference."

Galileo held up the camera, and took a picture, with the split
sensor, of what was now on the screen before them. It was an Italian word, SONO, the word for "I am," and SONO was seamlessly
displayed.

Indeed, said Galileo to J. As long as the sensor is in place, nothing will change, because every one of the million photodiodes will
go on reporting its own separate dot, unaware of what its peers are
seeing.

But what if, with a thin and sharp blade, one were to split in two a brain? Recall Ishmael's brain, in the crypt of Prince Venosa, when Salerno froze the connections between his two hemispheres. Would nothing change, as with the camera?

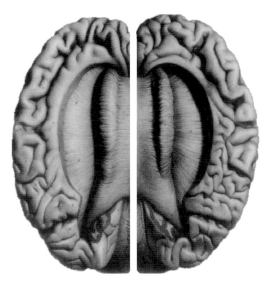

You know it already, said Galileo without waiting for an answer. Ishmael split into Ishma and El, and Ishma saw the lady, and El the brute, but no one saw them both—there was no Ishmael who could see the adultery, as long as the two hemispheres were split. But when the hemispheres embraced again with warmth, there was Ishmael again, and Ishmael saw the couple joined.

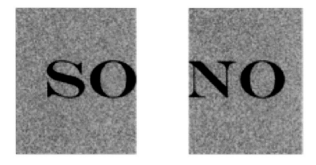

You know the answer then: Ishma would see *SO,* the Italian word for "I know," El would see *NO,* the word for "no," but there would be no Ishmael who would see *SONO,* and say, "I am." Unlike the camera image, the blade would split the conscious image and consciousness itself would be divided.

J. and Alturi remained silent, so Galileo went on. But if the connections between the hemispheres are warm, as they are in your own brain, you can try as hard as you may wish to split your experience in two, to see *SO* independently of *NO,* but you will not succeed. Just as you will not succeed in seeing the shape of things without their color, or their color without their shape—you will remain one J., one experience, one consciousness.

"Impregnable logic," said J. "One plus one equals two, but not quite," he added.

One thing is certain, said Galileo: there is nothing it is like to be the sensor of a camera—consciousness cannot live there, because the sensor is not a single entity, though it may be rich with a million photodiodes. Just like there is nothing it is like to be two scientists, one in the northern and one in the southern hemisphere. Nothing it is like to be a row of twelve men, each thinking of a different letter.

"I see it," said J. "The camera may be large, but is less than poor in consciousness: it owns none and lacks existence in the realm of experience. Compared to it, even a photodiode is richer, it owns a wisp of consciousness, the dimmest of experiences, one bit, because each of its states is one of two, not one of trillions. Yet being a photodiode is more than not being at all. I wonder," J. went on. "What if one splits the brain into a million parts? First left and right, than front and back into four quarters, then with a hundred other cuts through its white matter, into a million separate grains, as separate as the grains on a cob, or the photodiodes on the camera sensor: Would consciousness disintegrate?"

"Never mind," said Alturi. "Galileo hasn't shown a difference in number. If consciousness lives on information, one must squeeze money out of a formula, the formula of S."

Allow me, said Galileo at once, without raising his eyes. If we cut the camera sensor into its one million parts, the array of photodiodes, how much information is generated by each photodiode?

"One bit, of course," answered Alturi. "That's what the formula of S. tells us."

Now, said Galileo, how much information is generated by the camera sensor?

"What a question," said Alturi. "Being constituted of a million photodiodes, it will generate one million bits."

All right, said Galileo. How much information is generated by the camera sensor above and beyond its parts? Beyond its one million photodiodes, I mean.

"Zero, of course," said Alturi after a while, not expecting he would be questioned this way.

Precisely, said Galileo, feeling he was usurping Alturi's role. The camera does not generate any more information than the sum of its parts. Therefore, at least with respect to information, we have no need to invoke the camera above its parts. We might as well drop it from the catalog of useful entities, cut it with Occam's razor, and stick with a million photodiodes. *Entia non sunt multiplicanda praeter necessitatem.*

"That's just a matter of perspective," intervened Alturi, who seemed busy tipping the smoldering tobacco in his pipe onto the floor. "You like to talk of photodiodes and ban the camera, I may prefer the camera and spurn the photodiodes."

Not so, not so, hurried Galileo, think of Ishmael. He would have seen SONO and understood "I am." But after the connections between his two brains were frozen, and Ishmael had disappeared, nobody would be left who could see and understand SONO, "I am." Ishma and El together could never make up for it, for one saw SO and understood "I know," the other saw NO and understood "No." In this case, unlike with the camera, the whole is more than the sum of its parts and cannot be reduced to them; Ishmael is more than Ishma and El, and SONO cannot be reduced to SO and NO.

"I think I see the point," said J. "The information generated by the whole above and beyond its parts—call it integrated information—is what distinguishes Ishmael from a camera. Does this seem right, Alturi?"

CONTEXT

"What would be right?" exclaimed Alturi, who was busy grounding the tobacco out with his heel. "Is it right that a distribution of system states, if it cannot be factorized into a product of distributions of its parts, is not reducible? Of course, but SO what? What's special about this? There are all kind of things that cannot be factorized, that cannot split without a loss, but why would any of this matter for consciousness? Besides, there are many ways to divide a system into parts, of factorizing distributions, and you will get a different answer depending how you cut it into pieces."

"True," said J. "If integrated information has something to do with consciousness, it should not change depending how you divide a system into parts. Isn't it so, Galileo?"

The crucial cut, said Galileo, is the minimum cut, the cruelest cut of all—the cut through a system's weakest link, the cut that divides it into its strongest parts—those that generate as much information as possible by themselves, leaving as little as possible for the whole.

"Excellent," said J. "Integrated information is the information generated by a system above its parts, where the parts are those that, taken independently, generate the most information. Now that we have a definition, we need a symbol for it."

"If you need a symbol, it should be Φ," said Alturi. "That is the symbol of the golden ratio—the right way of dividing something into parts. And the minimum cut, which reveals how much information is integrated information, is the right way of dividing a system into parts, is it not? You should call it Φ."

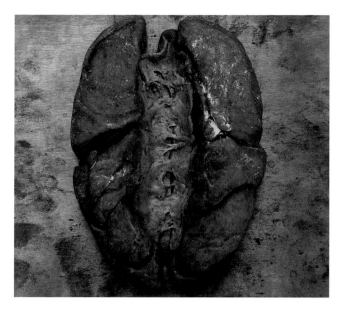

That would be interesting, said Galileo. After all, the golden ratio was studied by a fellow Pisan, the good old Φibonacci.

"It is better than that," said J. "Φ is like Φenomenology, like experience, which is what consciousness is."

Better than that, said Galileo. Φ has an I, for information, and an O, a circle, for integration. Let's call it Φ, then.

"Splendid," said Alturi. "Now that you have your quantity and your symbol, let's see what follows. Clearly, every time some elements interact, you'll have some integrated information: a whole that does not reduce to its parts. Then, if integrated information has something to do with consciousness, as you seem to think, what follows is quite simple: it follows that consciousness is like an onion.

"Take me and the neurons in my brain. Somewhere inside my brain there is me, of course, but I am not alone. If you peel me away, neuron by neuron, you'll find other me, millions of me, each lacking some part, but all conscious to some extent. I am just the most conscious of my many me, but those diminished selves would be right in claiming their own rights, except that I don't hear them, but they are along for the ride.

"Then take my body. No doubt my body, too, is made of interacting parts, a whole that cannot be reduced to its parts—either physically or informationally—in fact the brain itself is just one of those parts. So the body, too, is yet another consciousness, an even larger onion than I thought I was. Its Φ may be much less than mine—its minimum cut quite weak, say across my neck—but it, too, is carrying on its own limited existence. A multiplication of selves, a proliferation, of which I know nothing at all, and they know nothing of me.

"But it does not stop there. Then there is the two of us talking, nay, the three of us, interacting as a whole that cannot be reduced merely to the three of us taken independently. A Holy Trinity thinking its little Trinitarian thoughts.

"And then a city, a country, or the entire world, all layers of the universal onion, and each of them conscious, some more and some less."

"I see your point," said J. "And yet consciousness seems to reside just once inside my head, your head, and Galileo's head. Then Φ cannot be the answer."

The onion, too, must be peeled with Occam's razor, said Galileo. And when you do so, that will leave only its core—the core where integrated information reaches its maximum—the core that holds together while the rest peels off.

"So consciousness is not an onion, it is an onion's core! This is quite some progress," said Alturi. "But if you and I talk, what then? Don't you and I, talking as we are doing now, form a larger core?"

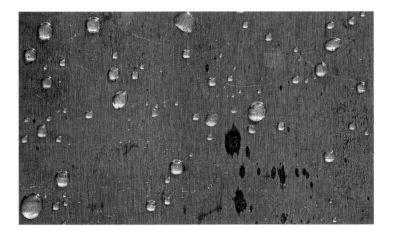

Occam's razor, once again, answered Galileo. You just said: "You and I, talking." "You and I, talking" is much simpler, physically or informationally, than a would-be chimera mingling you and I. That monster has no holding power and would break down at its seams, you and me, where reality is carved into individual entities. Think not of monsters but of raindrops. Inside a drop of rain, molecules interact more strongly than with the air outside, and so a surface forms. The drop is a single entity and is contained within a border. When two droplets meet, either they bounce and remain separate, or they fuse and become a single, larger drop. There are no overlaps, nor drops within other drops. So it may be with consciousness: consciousness lives within a system where integrated information reaches a maximum, inside its own drop.

"So what you have understood is this," said Alturi: "Experience cannot be reduced to anything less then it is. Impressive indeed."

Ignoring Alturi, J. turned to Galileo. "If you are right, we should have a name for a system for which the information generated by the whole above its parts reaches a maximum, the onion core, the raindrop of consciousness. A complex, perhaps?"

Let's call it so, said Galileo—a complex.

"So a complex is where consciousness lives," said J. "There consciousness raises its house, erects its walls, and you are what's inside, the rest of the world is what's outside. The house of consciousness is one and cannot be shared: there is only one, only one owner, and it excludes all others."

It was not clear whether Alturi liked this, but then he said: "I guess when you apply this analysis to the sensor of the camera, it will break down into complexes that are individual photodiodes, each of them distinguishing between just two states, on or off, but there will be no integrated entity—a complex—corresponding to the sensor. But when you analyze your brain, you will find inside it a set of nerve cells that form a large complex: one that can distinguish among a large repertoire of states in a way that its parts cannot; and one that does so maximally, more than any other set of nerve cells, more than the entire body, than any crowd of men, than the world itself."

Precisely, said Galileo.

"Then I have something for you," said Alturi, and handed Galileo some notes. The notes were from Frick and were full of diagrams representing parts of the brain. There was the cerebrum: without it, Copernicus had lost his consciousness forever. Galileo remembered when he and Frick had compared the cortex and thalamus to a great city. The diagram showed that a large expanse of the cerebral system formed a single complex of high Φ. This was because its elements, different groups of neurons, were specialized for different functions, and yet these specialists talked to each other—they were integrated within a single great complex that could distinguish among a vast number of different states, one for each experience.

There was the cerebellum, which had even more elements, but they were separated into many small modules that did not talk to each other. Each of them formed a small, separate complex, and for each little complex Φ was low. Like a collection of photodiodes, thought Galileo, and remembered Poussin: the painter's hand trembled without a cerebellum, but his mind was rich and full.

Then there were diagrams explaining why your eyes may be blind but your consciousness can have inner vision, like the blind painter in front of his great allegory. They showed how the visual inputs reached the cerebral cortex, influenced its functioning, but did not become part of the great complex of high Φ that gave rise to consciousness.

And there was his friend M., too, showing that all the nerves reaching out of the great complex, though necessary to speak and act, did not participate in it and thus did not contribute to his consciousness. There were Galileo's muses, the poetess and the gamba player, with loops going out and into the great complex, but the loops themselves remained outside. This was why, thought Galileo, so many neural processes that make us understand speech, or find the right words, or say them, or remember them, perform marvelous feats, but still remain outside the special sphere of consciousness.

And finally there was Ishmael, with the nerve fibers linking the right and left hemisphere split, and the great complex splitting in two and yielding two consciousnesses with similar values of Φ, Ishma and El. Other, smaller splits, might explain why Teresa could see and yet did not know it, thought Galileo. And surely Φ was low during the frenzy of seizures, or the deep waves of unconscious sleep, because the repertoire of distinguishable brain states was bound to shrink.

"You think this might explain it?" asked J. after a while.

Consciousness is such a lofty bird that it must be caught with an equation, said Galileo. First catch the concept, then dress it in the language of mathematics. Then, and only then, knowing how it can be measured, would one truly know what it is. It may be, said Galileo,

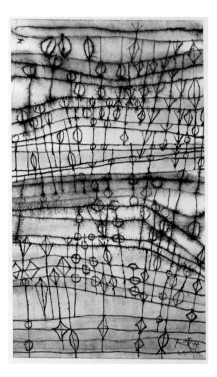

it may be that the essence of consciousness is integrated information. And this may be a way this concept can be grasped—a way to catch this bird: a way to find what entity is a single entity, a nucleus of experience.

"Something still perplexes me," J. said pensively. "The brain is inconceivably complex, so much so that trying to understand its mysteries through a network of equations is like trying to collect the sea with fishing nets. The brain has more trees than the jungle, more streets than a great city, is more plastic than the desert's sand, more changeable than the waves of the sea. And who would hope to reduce the endless wavering of the dunes, the bustling traffic of the market, the tangle of leaves and animals in the jungle canopy, to a series of equations, or worse, to a set of numbers? Mathematicians may weave their networks, but in the end, I am afraid, they will catch nothing."

"Do not be afraid, since there is beauty in principles," said Alturi's voice from a distance, and J. turned to question Galileo.

But Galileo too was far away. For he had read something at the end of Frick's notes, something he recognized from long before:

Philosophy is written in this grand book—the universe I say—that is
wide open in front of our eyes. But the book cannot be understood unless
we first learn to understand the language, and know the characters, in
which it is written. It is written in the language of mathematics, and
its characters are triangles, circles, and other geometric figures, without
which it is humanly impossible to understand a single word of it; without
these it is like wandering in vain in an obscure labyrinth.

So Galileo felt, for the first time in a long time, that he must write
what he had learned. And this is what he wrote:

Integrated information measures how much can be distinguished by the whole above and beyond its parts, and Φ is its symbol. A complex is where Φ reaches its maximum, and therein lives one consciousness—a single entity of experience.

NOTES

"The whole is more than the sum of its parts" is an expression that comes from Aristotle's *Metaphysics,* which Galileo knew well. William James thought that integration was a key to consciousness and fought hard to understand it, as revealed by some excerpts from his *Principles of Psychology* integrated into this chapter. Unfortunately he never succeeded and eventually gave up amid doubts and denial, writing an essay with the revealing title "Does Consciousness Exist?" James's photograph is from the Houghton Library, Harvard University. The all-sky night photograph of the Northern Galactic Hemisphere (on the left) was taken by Tunç Tezel at the Canary Islands; the Southern Galactic hemisphere (on the right) was taken by Stéphane Guisard in the Acatama Desert. The picture obtained by joining the horizons of the two all-sky images was the Astronomy Picture of the Day on July 30, 2011 (A Tale of Two Hemispheres).

If integrated information, measured by Φ (the Greek letter phi), is indeed the weighty concept at the heart of consciousness that it is claimed to be, this chapter introduces it in a rather light-weighted manner. Perhaps the author was trying to avoid equations at all costs, but the result is far from satisfactory. Versions of Φ dressed in equations, but in the end just as unsatisfactory, are found in Tononi and Sporns, *BMC Neuroscience* (2003); Tononi, *BMC Neuroscience* (2004); *Biological Bulletin* (2008); Balduzzi and Tononi, *PLoS Computational Biology* (2008); Tononi, *Archives italiennes de biologie* (2010, 2011). Information was defined as "a difference that makes a difference" by Gregory Bateson, in *Steps to an Ecology of Mind* (University of Chicago Press, 1972). *The Ripe Harvest* by Klee is at the Sprengel Museum, Hanover, Germany. The last portrait is by Arcimboldo (disliked by Galileo and modified in bad taste, if not bad faith) and is known as Adam (Eve's counterpart, private collection).

17

GALILEO AND THE BAT

In which is feared that the quality of experience
cannot be derived from matter

No explorer can resist a cave—promise of depth and secrets. So Galileo had penetrated deep into the cave, sure it would guard a revelation. But eagerness had lured him too far inside, and now he stood, uncertain where to go, his purpose lost, nothing for him to see save his own blindness. And then the fear began. He imagined dark shadows trembling on distant walls; his fingers felt the rock's wet face, scared it would break loose. His foot slid over the brittle edge, and his heart plunged toward the abyss, bolting from the weak hold of his will.

Then came the flutter. Suddenly, from the unfathomable heights over his shoulders, came the flutter, and it was made less of sound than of cold air that briefly touched his brow and froze his sweat. It was a flutter like the flutter of a bat, but not an ordinary bat. As if a giant vulture would fly as swiftly as a bat. Each time the flutter came closer, and it was chilling.

The cave had been violated, the bat was certain now. Finally it had happened, he thought. He had been careless, loitering in the air without a purpose, wasting time in useless dances, when he should have built ramparts and traps. And now it was too late.

The intruder must be of hefty size, perhaps extraordinary size—not one of the small crunchers that on and off disturb the quiet of the night but can be silenced in a sweep. And its echo was confusing, thought the bat—perhaps it hid behind the seventh bridge, and from behind the bridge he felt the deafening heat. He should have built stronger defenses, he thought, collapsing arches and piercing entrapments. But now it was too late. The intruder had trampled over all his trenches, and it was breathing heavily, not far from the inner chamber.

The bat swept again close to the bridge, trying to echo the intruder from a safe distance. Then, turning swiftly in midair, he felt a sudden *twong*, a *twong* of unbearable intensity. What was this twong, a twong so concentrated and powerful it filled his mouth with juices? A *twong* that was drawing him, irresistibly, to an aspiration wider and deeper, fuller and heavier than any before?

The bat quivered with dread. The heat behind the bridge was so loud, there could not be any hope to aspirate the intruder. He heard again the *twong*, drawing him closer. Ah, thought the bat, despite the *twong*—despite the *twong* he must resist. Perhaps the intruder was so powerful that it could aspirate a cruncher twice his size, maybe ten times his size. Perhaps it was so cunning that it could shrike him onto his own impaling traps. Certainly it had come to turn his chamber into a grave—waiting behind the bridge, only to show unholy force at the last instant. So he would fall, doomed by his sloth, thought the bat.

And then he thought, at least he'll fall with a *twong* filling his mind, a *twong* so intense it would occupy it fully and leave no room for dread. Perhaps that was the way to fall. He felt his wings fluttering irresponsibly. He felt juices spilling from the fissures in his mouth, and the echo became distorted, enlarged, out of focus. The heat was growing louder, and nearer, and his body was ready to swing, and then again he felt it—the *twong*, the strongest, most acute *twong* he had known, and he flew himself headlong, at magnificent speed, with a wide, arched swoop, behind the shadow of the seventh bridge, toward the roaring of the heat.

.   .   .

Galileo shuddered when he returned to his senses. A giant bat lay there, lifeless, his secret buried inside its skull. But voices were approaching, hunters perhaps, and they were arguing. Behind the rock, they could not see Galileo.

"Just what I meant," said a man named N. "We shall never know what a *twong* feels like. No matter how passionately you ask a bat, how assiduously you study its brain. Does a vulture hate the smell of carrion? And what do moray eels think when they, like sphinxes, stare unmoved out of their rocky holes, like question marks of stone?"

"How so sure, man of little faith?" a voice replied from the darkness. It could have been Frick's voice—but there was no Frick.

"Because," continued N., "even if we knew for sure that a system is conscious, we would never know what kind of consciousness it has. Because when experience comes, it comes in qualities that cannot be reduced to anything else. A flash of light is different, irreducibly, from the ring of a bell, or from twinges of pain; the color of the sky is different, irreducibly, from the shape of the sun. Nothing can explain why colors look the way they do, why red is red and blue is blue, why colors look different from shapes, and different from the way music sounds, or pain feels. Nothing can explain why a *twong* feels like a *twong*. And finally," said N., "because nothing that is made of matter, nothing, by any stretch of imagination, can hope to explain the quality of mind."

"Never mind matter," replied the other voice, "mind the facts. Mind that damage to certain parts of the cerebral cortex forever eliminates our ability to see, but not to hear, or to feel pain. That damage to other parts eliminates our ability to hear, but not to see, and yet other parts, if damaged, only affect pain. And mind that within the visual regions of the cortex, certain parts are necessary to see shapes, but not colors, others to perceive colors and not shapes."

Now Galileo could see N.'s head, and it was shaking politely. "I concede it freely. There may be a special relationship between certain areas of the brain and certain qualities of our experience," he said. "But that relationship is and will remain imperscrutable, just like every other relationship between matter and mind. Each of us lives in a cave without openings, just like the bat."

"Wait until a piece of the matter of your brain turns to sludge, and you will gain a better perspective on matter and mind," said the voice. "I met a painter once who had lost a spoonful of brain porridge, one particular spoonful. All was fine with him, except without that spoonful everything to him was gray and dirty, and he couldn't see, remember, imagine, or even dream of color. He saw food gray on the plate, his wife gray on the bed. But for the rest, his consciousness was just like mine—he had merely lost the special qualities of color, qualia, as you philosophers call them, and that was due to a dead piece of porridge. If a slightly different spoonful had been damaged, he would have lost faces instead of colors, or sounds, or the urgency of ethics. So there is something about how different areas of the brain are organized that makes them contribute different qualities to consciousness—sounds and sights, smells and pains, shapes and colors."

"That may be true, but still it is no use. No miracle will ever distill the bright red wine of consciousness out of the gray water of the brain," said N.

"Go ahead, revel in darkness as pigs revel in mud," said the voice. "Imperscrutable, how did it sound when you were born? How did red wine taste, when all you had experienced was milk and water? Surely you are a connoisseur by now—your sensations developed and refined—you are a philosopher. But that is just the point—development and refinement, how did they come about? There is no mystery in nature's book: they came from a rearrangement of connections among certain neurons in your brain."

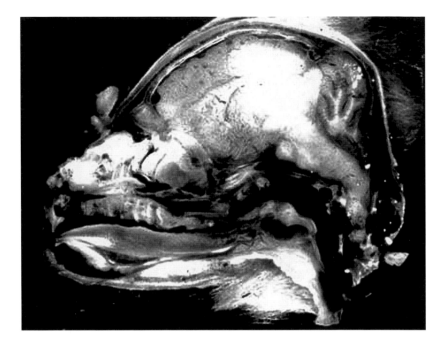

"There is no knowing what it is like to be a pig, or I would say you are pigheaded," said N. affably. "Fortunately that is philosophically impossible, or at least imperscrutable. Think of the poor bat. Assuming it was conscious, how did it experience the world it sensed through its sonar, sniffing the echo of things? Was its experience of the world more like a vision, was it instead soundlike, or was it completely alien? And what is a *twong* like? Is it like a twinge, or like a twang? Maybe in between, or completely different? Nothing you'll say will ever explain what it is like to be a bat. To feel a *twong*. No, nothing in matter can explain the quality of mind.

"And I shall add," N. went on, "if you don't care for the bat, think of yourself instead. Think of darkness, how it feels, think of a pain, then of the sound of water falling. Why should darkness feel exactly the way it feels and not feel otherwise? Why should it not feel like a bright blue sky? Why shouldn't blue feel green instead? Or feel like pain, and pain feel like darkness? Why should the fragrance of fresh bread not feel like the pangs of shame for not knowing the answer? Why, for that matter, should it not feel like a *twong*?"

Galileo lost track of the voices and could not hear the response to what N. had said. But Galileo was not going to accept that the kind of consciousness a system has is arbitrary—he was not going to abandon his old friend, the principle of sufficient reason—there must be a reason for anything to be the way it is and not another way. Just as the presence of consciousness depends on the functioning of the brain, so does its quality. That was the second problem of consciousness: What determines the specific way consciousness is? There must be something—some necessary and sufficient conditions—that determine exactly what kind of experience one has, thought Galileo, and the moment those conditions are understood, therein lies the explanation.

What was this something? It could not be some material attribute of a particular piece of brain—there was nothing red about the particular nerve cells that were necessary for perceiving red, nothing blue about those other cells that must turn on for us to see blue. Perhaps the explanation was to be sought at a different level. Whether consciousness was present, and where it was generated, he had come to think, was determined not by any property of neural cells but by the quantity of integrated information generated by a complex of neural clements. Then perhaps the specific way consciousness is—its *quality*—was determined not by any property of cells within the complex, but by the specific way the information was generated. Or so he thought.

NOTES

This chapter is sadly underdeveloped: a halfhearted attempt at consciousness's second problem, as confused as Galileo in the cave and logically as jerky as the flight of the bat. Yes, one gets the double perspective in Plato's cave, Galileo's frightened but familiar consciousness on one side, the heroic bat's alien consciousness on the other, but then there is a cacophony of voices that float in midair but offer no clear answer or solution. The problem, though, is obvious enough: anybody who has ever faced a moray eel staring back from an undersea hole in enigmatic immobility, like an old sphinx posing a metaphysical interrogative, will have wondered what his counterpart might be experiencing, and what it might be thinking. A bat is also a case in point, although it would be unlikely to stare back at us with philosophic poise. While morays would have been a better choice, bats have an illustrious pedigree: the philosopher Thomas Nagel, in the essay "What Is It Like to Be a Bat?" (*Philosophical Review,* 1974), chose such a creature, perhaps neither bird nor mammal, to make the case that the quality of consciousness will lie forever beyond the realm of science. Clearly all the *twongs* recurring in the chapter—apparently an all-important concept for the bat but a meaningless word for us—are meant to illustrate the unbridgeable gap between bat experiences and human experiences. A description of a painter who had lost the perception of color due to a localized brain lesion (achromatopsia) is given by Oliver Sacks in *An Anthropologist on Mars* (Vintage, 1996). Distilling the red wine of consciousness out of the brain's gray water was ruled out by Colin McGinn. *Plato's Cave* is by Kenneth Eward (1998). *The Boy Exposing a Bat to the Flame* is by Trophime Bigot, at the

Galleria Doria Pamphilj in Rome. *The Bat* by Albrecht Dürer is at the Musée des Beaux-Arts et d'Archéologie, Besançon, France. The green and greyface moray eels sharing a lair are from Green Island, South West Rocks, New South Wales, by Richard Ling. The brain of a vampire bat is from K. P. Bhatnagar, *Brazilian Journal of Biology* (2008).

18

Seeing Dark (Deconstructing Darkness)

*In which is said that darkness does not exist in a void
but requires a context*

All was dark, dark only, alone, lying in dark silence, and all was peace. Galileo stared at the night sky, the way he used to do as a young man. His mind was free: there was just dark, dark extending everywhere, dark as far as he could see. Nothing else—no moon, no stars, no planets. Unwittingly, slowly, his mouth murmured the very word: *dark*.

But Galileo's peace was brief. Dark in its purity was shattered, pierced by questions and doubts. What was going on in his brain when he saw dark? Which arguments among neurons were responsible for his experience? Could he now picture it, envision his brain and see what seed was needed for experience to be born? Could he now understand how dark—pure dark—truly came about?

A map is but a scanty thing. Colored maps instead of continents, mountains flattened, great cities lost but for their name, measure replacing motion, and temperature heat, all detail gone, no life inside. The world exceeds all description, events are far too dense for history.

Of all the maps, of all the models, that of the brain—even one crafted by a master of miniatures, full of intricate features—was a model twice dead. Dead because it was merely a model—a weary simulacrum of reality, and dead because it was inanimate, yet what it modeled had a soul. Even if its gears had cranked to life, its soul would have been missing.

Galileo was studying it anew, the shiny dome of the brain, trying to peek inside, consulting the books Frick had left him. What had he learned? Had he understood what it means to see? How an inner light penetrates the castle of the brain? His hand slid over its regions and felt the ceaseless whisper of neurons on the inside.

So he pictured a wave of neural firing, starting at the back of the eye and traveling all the way up to the highest visual areas of the cortex, near the top of the brain. There, as Frick had told him, the wave of firing induced by the dark sky would trigger a final burst of firing in a few select cells, the very ones that were specialized to signal darkness. Those were the cells that fired whenever we saw dark consciously, even if we just dreamed of dark, that never fired when we did not see dark, and without which—if they were damaged—we couldn't even imagine what dark looked like.

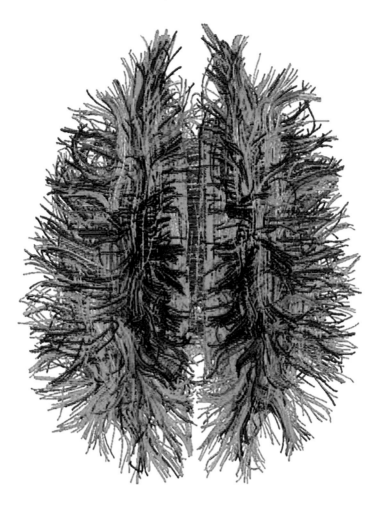

Galileo pictured the cells nearby, which were specialized to fire for light rather than for dark: they were completely silent, because there was no light to see. Other cells, not far away, which signaled blue- or red- or green-colored patches, those were silent, too, because all there was to see was an immense dark expanse. And cells detecting simple things, such as edges or bars, some vertical, some horizontal, and some oblique, some for edges in the center of the field of view, some to the right or left, they remained silent, too, since what they liked was absent. So were cells elsewhere in visual areas, cells specialized for detecting shapes, such as rectangles or circles, because there were no shapes to be seen. So were cells specialized in detecting faces, because there were no faces, or cells specialized in detecting objects that move, because nothing moved. And in other parts of the cerebral cortex, too, there was but little activity; there was no burst of firing in the auditory cortex, because there was nothing to hear, none in parts of the brain specialized for touching or tasting or smelling, because there was nothing to touch, taste, or smell, nothing to feel and nothing to think of.

Finally he pictured how these neuronal elements, distributed across many areas of the brain, were linked by myriads of interconnected mechanisms and formed a great complex: a single, integrated entity that worked as a whole, and it was maximally irreducible—a complex of high Φ.

Perhaps he could see it happen now: when he, Galileo, had seen pure darkness, the intricate mechanisms of this large complex had ruled out, in various combinations, trillions of possible states that could not have led to its present firing pattern—all neurons silent save for a few excited by darkness—and in ruling out those incompatible states, they had generated a large amount of integrated information.

Now he understood, he thought: inside the cerebrum was a single, large complex of high Φ, but not so elsewhere in the brain. He pictured the buzzing neurons in the cerebellum, in the brainstem, the spinal cord, and many other places: this buzzing was immaterial to his consciousness, this fire never fired with the spark of his own experience, not because these other neurons were different or weaker but because they were not privileged, they were not part of the majestic complex that reigned over his brain.

So cells in the retina, though they too had been firing in the dark, did not contribute to his experience: they merely ensured that the right set of cells high up in the brain, the privileged dark members of the corticothalamic complex, would themselves fire. So many other cells, confined inside innumerable provincial circles, in the cortex and underneath, like those that had made him speak, unwittingly, the word *dark,* were forever excluded from entering his consciousness. And so the circuits in the cerebellum, the spinal cord, and the brainstem—the ones that maintained his posture, stabilized his gaze, and controlled his blood pressure—were active day and night, fatiguing under the burden of their ceaseless tasks.

But they would never be part of the vast cerebral complex, the great, resplendent castle: their lot broke down into small, dark hamlets, mired in local arguments, their Φ miserably low. They were outcasts, never to be admitted to the magnificent hall where the rich, harmonious society of the castle was assembled perpetually, where all important news was received and discussed, all plans were formulated, and all decisions made. Outside, traders and workers and serfs went on in scores dispatching their menial labors and heard none of the news, knew not the plans, and had no say; yet relentlessly they supported the castle, without them it would soon cease to exist, not even the simplest task could be carried out without the tireless diligence of the outcasts. So now and then a dignitary would stand up and approach a small door near his seat and lend his ear through a porthole, perhaps issue in return some hushed order that nobody else could hear: a messenger would rush forth and whip the assignment into completion, perhaps return to the closed door to report and hear perhaps another order. But never, never would the messenger, or the serfs, partake of what was happening in the great hall, no whiff of that ethereal society transpired; all the serfs knew were mere reflections, indirect ripples of the great sea.

When Galileo awoke from his reverie of understanding, he was not enlightened, but shivering in doubt. He saw it all again, in rapid sequence, he saw the brain, the vast cerebral complex within it, he saw its neurons firing, and complexes, and repertoires, and he saw Φ. But there was an objection, as fatal as it was simple. Surely all he had thought was important to consciousness, perhaps essential—integrated information, the complex of flickering neurons. Yet just as surely, why should any of this, a complex of flickering neurons, ever *give rise* to experience? Why should integrated information acquire a subjective side and *become* experience?

Worse, it seemed to him that N. was right: there was no way to conceive how his complex of flickering neurons, flickering in that particular way, would give rise to that particular experience of pure darkness. No magic could wring his complex of flickering neurons and squeeze the juice of darkness out of it. No alchemy could make darkness out of neurons and their mechanisms. No mechanism could give darkness its special visual quality, its unique black quale. There is nothing, N. had said, nothing that can explain why darkness looks the way it does, why dark is dark and light is light, why colors look different from shapes, and different from the way music sounds, or pain feels, or a *twong* feels like a *twong*—there is nothing in matter that can explain the quality of mind. It was a foolish leap of faith—a leap that explained nothing. It all had been in vain. Galileo, once more, had been fooled by his fondest wishes.

A fool he was, but if he was a fool, he was a proud one, and Galileo's pride burned and fired his wit. He saw the image of Salerno heating and cooling Ishmael's brain, and then he saw that it was not Ishmael's but Alturi's, and it was not Salerno but himself, Galileo, and he was chilling it slowly, piece by piece, region by region, node by node—he was slowly going to erode Alturi's brain, like Count Ugolino, gnawing at the skull of the Archbishop of Pisa.

Alturi was there, staring toward the night sky, and all was dark, entirely dark—just dark, dark extending everywhere. And then Galileo asked what would happen if he chilled the parts of Alturi's brain that let him hear sounds—would his experience of dark change at all? Alturi's answer had come swiftly:

"Not in the least. If there was silence before, there would still be silence. As long as there is no sound, how could I tell that I am deaf rather than there is no sound?"

That would be true if I had chilled the inside of your ears, Galileo heard himself say. But having chilled the very parts of the brain that are responsible for your experience of a sound, how would you know there is silence instead?

"Are the absence of sound and the absence of hearing not the same thing?" exclaimed Alturi.

Certainly not, Galileo had replied. Imagine I also chill those parts of your brain that let you smell, and those that let you taste flavors. Would anything change?

"Again, no," said Alturi. "There would be a difference only if there were an odor to smell, or something to taste."

So now I will chill those parts that allow you to see human faces.

"Dark is not a face," said Alturi.

And now also those parts that are necessary to see any kind of shape, and those that make you see objects in motion.

"Dark is just dark, not a shape."

And finally those nerve cells that are necessary to see blue and yellow and red and green, finished Galileo.

Alturi hesitated for an instant, and then he said that no, his perception of dark would stay the same. Of course, if a blue light had been turned on, an object had appeared, a face materialized, a bell had sounded, he would not have noticed, he would have just gone on seeing dark, but dark it was.

When you say you see dark with your brain mostly chilled, do you mean exactly the same dark as when you say you see dark now? was Galileo's next question.

"Exactly the same," said Alturi without hesitating.

But how would you know that dark is dark rather than a color, if you would not even know, anymore, what color means?

"Why would I need to know about color, if all I need to do is see dark?" said Alturi.

And how would you know that you are seeing rather than hearing something, if you cannot know what hearing means, nor smelling or tasting? insisted Galileo. But seeing that Alturi would not answer, he told him he would now proceed to chill, at last, those cells he needed to see light. He would erode Alturi's brain down to pure dark.

Alturi hesitated longer. "I would still be saying that I see dark, as long as the parts of my brain that allow me to speak are not chilled," he finally said.

And what would you mean, when you say that you see dark? You would not even know that dark is the opposite of light, let alone that it is not a color, not a visual object, not a sound, a smell, a taste, a thought, a wish, a memory, or a feeling, said Galileo. You would know hardly anything at all. And you would *be* hardly anything at all: you would generate one bit of information—a photodiode, that's what you would be like.

Staring at Alturi, who was silent, Galileo went on. This is what is difficult—the problem that is hard. When you see dark, you think you see just that—dark—you think you see something that comes already labeled, and all you need to do is watch it, to see it is dark. Darkness, you think, is something that's out there, its darkness ready made, and all you need to do is see it—you do not have to build it, compare it to anything else, to know indeed that it is dark. And certainly, to see it is dark, you do not have to say to yourself: what I see is not light, not blue, not red, not a ripe apple, not Pisa, Padua, or Rome, not a sound, not a feeling, this is none of the countless scenes I might experience. In fact, you need not go through any of those experiences, nor would you have enough time to do so, even if you were to count one each second and spend a thousand times your life doing so.

This is the trick that consciousness plays on you—your experience of pure dark is given immediately. But in reality it can be dark only in relation to what it's not. And it is only if you have a brain that can distinguish darkness from untold other states, from each in a specific way, that you actually see dark. When the photodiode tells dark, that dark is empty, merely one out of two. When you tell it's dark, it is one out of trillions, made what it is—special in its own special way—by how it differs from each and every one of the other trillions. Your dark is full—extraordinarily full—as full as the most bustling market in the world.

And then Alturi began to see, and Galileo with him. He saw under Salerno's hands his own cerebral complex and saw the neurons that, alone, were firing for dark, but they were not alone. The other, silent neurons of his cerebral complex, the quiet ones, they were required to give darkness its context, to put his black in context. Without those quiet ones, without their mechanisms spread out throughout the cerebrum, the darkness neurons were like humble photodiodes;

to them, taken alone, the world was also made of two, this way and not another way, but was not made of dark and light.

And then he saw Salerno loom over his complex and cut off its connections, shatter its mechanisms, first around the color cells high up, and when he did so, a small portion of the complex shrunk. And he saw his experience would shrink, too, his dark would become less dark, because he did not know anymore that dark lacked color.

And then Salerno was madly breaking mechanisms, eroding the repertoire that could be distinguished, though all the nodes were left in place, the dark one firing its burst of activity, and the others quiet but intact, all the connections severed. At that the complex collapsed and broke down into countless fragments: the great diamond had disintegrated. And then experience was shattered, consciousness was shattered, its quantity annihilated and its quality evaporated, and dark would be no more.

NOTES

Galileo imagines what his brain might be doing when he sees dark, in order to give concrete form to the intuitions he has developed about consciousness, starting with the photodiode thought experiment. Galileo's model of brain activity is extraordinarily crude, and his specific example questionable: it is not clear whether there are brain cells firing selectively for dark, though there are cells that fire in response to the overall brightness of a surface, and there certainly are cells that respond to a given color the way we do. For example, in visual areas V4/V8 of the cerebral cortex, there are cells that discharge every time we see blue and do not discharge when we do not see blue; if they are damaged, one cannot see blue, and one might add, if one electrically stimulated them, one would see blue even in complete darkness (which is not unlike what happens when we dream of blue, or imagine it). Such cells are the scientists' best guess as to the "neural correlate of blue." As rough as Galileo's understanding of brain activity may be, his conclusion seems reasonable: to see dark (or blue, or whatever) consciously, our brain must house a neural complex endowed with a large repertoire of distinguishable states—instead, a photodiode specialized to signal dark (or blue, or whatever) would not see at all, whether it is a piece of electronics or the equivalent circuit in our brain. The neural substrate of consciousness has been irreverently inserted into Vermeer's *The Astronomer* (Louvre). The network of connections in the brain, also known as a connectome, is courtesy of Olaf Sporns and Patric Hagmann. The dome is from the Haghia Sophia, Istanbul. Dürer's engraving *Melancholia I* contains a magic square, a mysterious polyhedron, and the instruments of geometry and reason. Gustave Doré's illustration of Ugolino's gnawing the head of Ruggieri is for Canto XXXII, *Inferno,* of Dante's *Divine Comedy.* The brain eroded is modified from Dr. Paul Thompson's images of Alzheimer's disease, UCLA. *Chicago Board of Trade II* is by Andreas Gursky. The diamond is from a private collection (Bridgeman).

19

The Meaning of Dark
(Constructing Darkness)

*In which is shown that darkness is built of many
nested mechanisms that specify what it is not*

Who paints night's darkness dark? How does its pure black juice ooze
out of mechanisms of matter? Why is darkness dark, but light is light,
colors don't smell like the scent of citrus, differing from sounds? And
why does pain not taste like wine?

This was the challenge N. had posed, but Galileo now had a clue.
By freezing one region of the brain after the other, he had seen darkness disappear. Every time the great cerebral complex had lost another
set of mechanisms, region by region, every time pure darkness—the
simplest of experiences—had lost its meaning, shade by shade. So
those other mechanisms were needed to make darkness what it was,
to give it context and thus to give it consciousness. When all mechanisms were frozen, all except for the neurons distinguishing dark
from light, just one bit of consciousness was left, and that single bit of
consciousness was left bereft of quality.

Then out of the darkness a scene emerged, and out of silence
sounds, of wheels turning without pause. Galileo could see them,
thousands of cog wheels, some pin small and some as large as a mill
wheel, that were revolving each at its own speed, like a deranged
clockwork. Some of the wheels were connected by thin shafts moving incessantly up and down. Other, more distant wheels were connected by long chains tied around their sprockets, making elaborate
turns around cantilevered pinions. There seemed to be no end to the
revolving gears, a million ropes crossed each other in all directions, as
far as the eye could see.

Deafened by the noise, Galileo could not bring his mind to rest.
There was no way to uncover what the elaborate machine might do,

save cranking its countless wheels to the end: he could not see anything going into it, or anything coming out—it did not seem to serve any meaningful purpose.

But somebody came up to Galileo, addressing him with living force and a faint German accent. "You will admit, esteemed colleague," he said, without expecting a reply, "that experience cannot be explained in terms of *causatio mechanica*—that is, *salva veritate,* in terms of shapes and motions. What you see is a giant machine, which was constructed to give rise to thought, sense, and perception. You can investigate every part of it—its size is such that you can walk inside it, like going into a mill, but you can wander long and wide, and all you'll see are working parts pushing each other—you'll never see any thing or soul that would explain experience.

"Look there," the man said, and took Galileo behind a long stone wall. There were a multitude of people, dressed in the Oriental manner. Each was standing on a small pedestal, peering into a telescope, and each one held a flag, either up or down.

"Calculemus!" said the man. "Cogs can calculate, pulling chains and ropes. Ants in an anthill calculate, if only they knew what. Millions of Chinese men can calculate, waving flags to each other. Calculations can be mechanized. Thought can be mechanized. But can experience be mechanized, Herr Galileo? What is it like *to be* my giant mill? Is it like anything at all? Ach," he then added, "you must forgive my rudeness." He bowed elegantly. "I should have introduced myself, Baron von L., here to serve you, my most acclaimed guest."

Most honored, Baron, said Galileo. As to the mill, if I remember well, I'll answer the way Bruno has done (Bruno, who had not abjured, thought Galileo). To a visitor, we too are cog wheels:

> *They live like worms within an animal, all the animals within the World, nor do they think that the World feels anything, just as the worms in our belly do not think that we feel anything, and have a soul greater than theirs . . .*

"Poetry, my dear colleague, poetry," said the Baron, "but now it is time for logic, I found it the best of medicines! Where was I? Of course, there I was, *fiat lux!*" he added with a well-meaning smile, "if you don't mind some questions. So if I may, Herr Galileo, what have you learned from the photodiode, as it were, that you may graduate to the mill? The light is on, and the photodiode turns on, what does it know? *Quod noscet?*"

This is what I learned, said Galileo. The photodiode has no mechanism to distinguish colored from colorless light, even less to tell which color light might be. All light is the same to it, so "light" cannot possibly mean colorless as opposed to colored, let alone colored in a particular way. Nor does the photodiode have a mechanism to distinguish between a homogeneous light and a bright shape—any bright shape—on a darker background. So "light" cannot possibly mean full-field as opposed to a shape—any of countless particular shapes. Worse, Galileo went on, the photodiode does not even know that it is detecting something visual, since it has no mechanism to tell what's visual, such as light or dark, from what isn't, like hot and cold, light and heavy, or loud and soft.

"Ausgezeichnet," said the Baron. "As far as it knows, the photodiode might be a sensor for temperature rather than for light—it has no way of knowing whether it is sensing light *versus* dark or hot *versus* cold. As my eminent colleague surely understands, the only *specificatio* a photodiode can produce is whether things are this way or not, *sic aut non sic,* as it were. Any further *specificatio* is impossible, because it has no mechanisms for it. So when the photodiode detects 'light,' such 'light' cannot possibly mean what it means for Herr Galileo; it cannot even mean that light is a visual thing.

"But *you,* renowned colleague," continued the German, "when *you* see 'light,' you are wondrously specific, whether you know it or not: you have exquisite mechanisms in your mill, your *moulin merveilleux,* mechanisms specifying at once that things are this way and not another (light as opposed to dark), that what you are distinguishing is not colored (of any particular color), does not have a shape (any particular one), is visual as opposed to auditory or olfactory, sensory as opposed to thoughtlike, and so on. Each mechanism is a concept, Herr Galileo. Meaning from mechanism," exclaimed the Baron: "That's where the meaning of light and dark comes from!"

Galileo nodded. By realizing what the photodiode lacked, Galileo had begun to appreciate what allowed him to "see" the light.

"Most excellent," said the German, "but Herr Galileo surely will concede that so far he went the easy way, *per subtractionem,* as it were. He imagined subtracting one mechanism after the other from the brain, and saw that consciousness of 'light' would degrade progressively—would lose its noncoloredness, so to speak, its non-shapedness, would even lose its visualness—its meaning stripped down to just 'one of two ways,' as with the photodiode. That's very well indeed," he went on, "but now it's time to properly understand, *per additionem,* as it were. All those critics, Herr Galileo, they think they can demolish my edifice. Sure, demolishing is easy, but I can build faster than they can tear down, add faster than they can sub-tract! What was I saying, indeed? As a matter of fact, esteemed friend, can you make light from scratch?"

Galileo knew what the Baron meant. Which mechanisms would one have to add in order to augment the lowly photodiode and make it like Galileo's brain? Or, which really was the same, to augment the neurons in his own brain that could tell light from dark, and nothing more? One would need to add many other mechanisms, which could distinguish other things, colors and shapes and sounds. So "light" would mean what it meant and would become conscious "light" *by virtue of* being not just the opposite of dark but also different from any color, any shape, any combination of colors and shapes, any frame of every possible movie, any sound, smell, thought, and so on.

"Most exactly," said the Baron, "as long as distinguishing 'light' from its alternatives is not picking one thing out of an undifferentiated bunch, as it were, but discriminating in a specific way between each and every alternative—*specificatio plurima!* And now let us imagine," he exclaimed. *"Imaginemus!"*

So Galileo tried to imagine it: he imagined a simple mechanism, like that of the photodiode, somewhere in his brain, which could tell light from dark in the center of his vision, and the mechanism was built with neurons and connections. That little mechanism generated some information—just a bit of information, though by itself it had no idea what light was, or what dark was, or that it was in the center of vision. Then he imagined another simple mechanism that could tell blue from other colors, nothing more, again in the center of vision. That mechanism, too, generated a small amount of information, telling apart blue from not blue at the center. And then another mechanism that could tell red from not red.

"Most brilliant," approved the Baron. "And then of course there are similar mechanisms, Herr Galileo, other neurons and connections that look at a different part of the visual field, a set just slightly right of center, another left of center, and above, and below, and further right and further left. Millions of such little mechanisms, repeated for light and dark, blue, red, and so on."

And then again, said Galileo, there will be other neurons, standing on top of this first layer, whose mechanisms calculate further, based on what some of the lower neurons do. For instance, one of the higher neurons might turn on if three of the lower neurons do so in a row. Such a neuron would distinguish between the presence and the absence of a little horizontal bar at the center of vision.

"Though of course," added the Baron, "the neuron by itself would have no idea what a bar is, or what light is, or what the center is, just like the photodiode."

Yes, said Galileo. Still, there would be many more such neurons, not just for horizontal bars but for vertical and oblique ones, and for different parts of the field of view. Then higher up there would be other neurons that can detect, by combining what lower neurons do, the presence or the absence of particular shapes, or maybe faces or places, and so on.

"Indeed, indeed," said the Baron. "And lo and behold, very esteemed colleague, some of these neurons might even achieve a feat Plato thought was magical—some of these neurons might distinguish an idea! The idea of a triangle, wherever it may be, no matter how large or small, no matter where its corners are pointing, no matter whether equilateral, isosceles, or scalene!"

Yes, said Galileo, or an idea like a human face, perhaps *her* face, no matter where it might appear—center, left, or right; no matter which way her gaze may turn, no matter how she is dressed, whether she is smiling or sad, in person, in memory, in a dream.

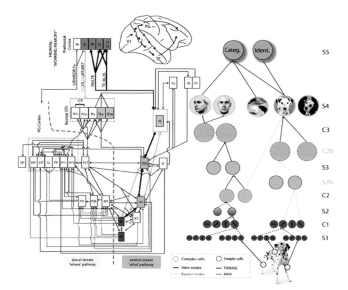

And then . . . the Baron interrupted Galileo's reverie. "My dear friend," he said, "before you are flown away by your imagination, let's not forget that each of your little mechanisms, no matter whether lower or higher ones, just does its little thing, the only trick it knows, without knowing more than a photodiode does. I mean, there may well be a neuron in my esteemed colleague's brain, distinguishing this feminine entity, or should I say the idea of this feminine entity—*das ewig Weibliche,* as it were. But while this neuron may proudly grasp the idea, *ipso facto* it will forgo the details; because the idea, the invariant, the universal, is what it is precisely by abstraction, by regally ignoring the particulars."

The particulars, thought Galileo—where she was, where her eyes turned, her dress, her colors, and if her smile was sweet—that was the task of other neurons. Galileo knew it by now.

"Très bon, mon revered *collègue,"* confirmed the Baron. "Each neuron, each little mechanism, rules in its tiny kingdom—rules out some possibilities, specifies its small domain, its tiny concept, but it knows nothing of the rest. From the outside, we may discover a mechanism for detecting light in the center, another one for light on the left side, one for blue and one for red, one for oval and one for square shapes; one for noses, one for lips, and one for faces, and maybe even one for her, whoever she might be. But the little mechanisms, they do not know it! For each of them, it's merely this or not this, *sic aut non sic,* their individual universe is rounded with one bit."

This is now clear to me, said Galileo. But then for consciousness to emerge, these mechanisms must work together. They must specify together, at once, many nested concepts: that it is her, that she is in the very center, she is wrapped in black, her eyes looking into my soul, her mouth curved with affection.

"As a matter of fact," said the Baron, "and what they specify together must be more than just the product of what they specify in isolation. The whole must be much more than the sum of its parts—otherwise there would not be any concept, and there would be no complex, no single entity, no her, and no Herr Galileo."

Yes, thought Galileo, but how? The mechanisms must be arranged in a certain way, such that they form a whole, a complex that lives above and beyond its parts. A mill was not enough, he thought. The cerebellum was as intricate as any calculating mill, yet if there was experience there, it must be a meager thing indeed: you could evict all cerebellum from its seat in the brain, and experience would flow as crowded as before. But the cerebrum, a mill neither more alive nor more endowed than its neighbor, bore the fruit of consciousness at its ripest: both lived inside the skull, they both were made of wholesome elements, both were enchanted, woven by nature's looms. But only a blow to the cerebrum would whiff the soul away: the intricacy of a maze of vines, the warmth of a soft build, being carried on the breath of life, were not enough to guarantee experience, just as being made of cogs did not preclude it.

"Yes," said the Baron, "what matters is the circumference of the possible, the repertoire of a single entity, closed and alone: and when you measure its cipher and know its Φ, the largest, most complicated web—made of Chinese or Christians, of cogs or cells, of flowing wires or beams of light, infinitely intricate to the eye—might decompose into a multitude of small complexes, each of them not worth much thought; or it might reveal itself a diamond with more faces than the great world—a cold-looking mill aflame with a burning soul."

You give me poetry, too, dear Baron, said Galileo, and now it's time for logic. Tell me, how can this be done?

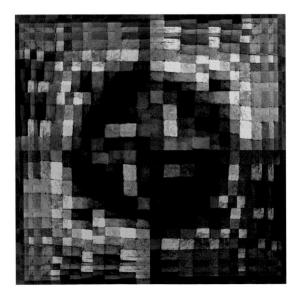

"Elementary, my dear Galileo!" exclaimed the Baron. "By piling one mechanism on top of another, that's how it's done! Pelion upon Ossa! One mechanism alone—things are this way and not another way—generates hardly any information: one nondescript bit. But if you pile them up, so that one builds upon another, not only side by side, but each standing on the shoulders of the others, as it were, they generate far more information, bits multiply, the nested concepts become more general, their unions more specific."

Perhaps, said Galileo. Perhaps if I could picture these nested concepts and ideas, how they weave together and join within a single structure, perhaps I could build meaning out of mechanism.

"Most honorable colleague," exclaimed the Baron, "you still dismiss my mill! You do not like its ropes and cog wheels? Perhaps you value a more ethereal kind, and you would rather play with light. Then go, and see my *alter ego*. But what are a mill's ropes, if not taut springs of gold? And what are cog wheels, if not the vertices of a precious diamond, and its stupendous geometry, the enchanted loom engendering the entangled web, the space of qualia—the space where feeling is defined by mechanism?"

NOTES

The German philosopher who entertains Galileo is obviously Leibniz (his portrait is at the Niedersachsisches Landesmuseum, Hanover). Leibniz imagined his mill thus: "Suppose there were a machine built to think, feel, and have perception, suppose it were increased in size, keeping the same proportions, so that one might go into it as into a mill. That being so, when examining its insides, we should find only parts which work one upon another, and never anything by which a perception could be explained" (*Monadology*, 1714). Later philosophers, such as Ned Block and John Searle, elaborated upon Leibniz's mill, devising Chinese variants of the argument. Unlike Block and Searle, Leibniz was a panpsychist and thought that all corporeal substances had soul, unless they were united in a merely mechanical manner (mere aggregates, such as a heap of rocks). In this, and in the use of the notion of monad, he resembled Bruno, as Galileo correctly remembers. Interestingly, a monad has no parts, yet each one is unique, and there are some dominant ones. Leibniz was trying to deal with the difficult combination problem that has always plagued panpsychism—if everything that exists has some mind or consciousness in it, even atoms (or, we would say now, subatomic particles), how is it that some things seem to have more consciousness than others—the cerebrum much more than the cerebellum, and incomparably more than a stone or an atom—and some aggregates may have none at all? If things are bad for panpsychism with respect to the presence of consciousness, they are worse when it comes to understanding the quality of experience. Which is exactly what Galileo and Leibniz are trying to do, building up a hierarchy of nested mechanisms in the brain and hoping to get somewhere by climbing into this modern version of the mill . . .

The print illustrating the mill is from Piranesi's *Carceri* (*Prisons,* Plate VII, "The Drawbridge"). The calculating machine was invented by Leibniz himself. The woman with the light is by Godfried Schalcken (Palazzo Pitti, Florence). The diagram illustrates the connectivity patterns in the visual cortex of the macaque monkey (left) and a hierarchical model of object recognition (right), adapted from G. Kreiman, "Biological Object Recognition," *Scholarpedia,* itself adapted from T. Serre, M. Kouh, C. Cadieu, U. Knoblich, G. Kreiman, T. Poggio, Artificial Intelligence memo, Massachusetts Institute of Technology (2005). The patchwork, presumably illustrating a matrix of "informational relationships" among the elements of a system, seems to have been liberally adapted from a painting by Klee (*Ancient Harmony,* Kunstmuseum, Basel).

20

THE PALACE OF LIGHT

*In which is shown that an experience
is a shape made of integrated information*

Before his eyes a faint grayness hovered, like vapors from a frozen lake before night falls.

Galileo heard a voice high above. "This is the source of ignorance absolute," it said. "This is where you take your start."

"Why ignorance?" asked another voice from far away.

Perhaps it was Alturi, hoped Galileo.

"You shall know soon enough," said the voice from the heights. "Walk until you find a rope, and be prepared to climb."

"So ignorance is like a gray and homogeneous fog—like a bad day in England," commented Alturi.

"Only through mechanism does ignorance retreat," replied the voice from the heights. "But hark: each victory is only a few bits' worth, nothing against the enormity of unawareness. Hence, do not linger."

So Galileo advanced in the mist of ignorance, fearing the uncertainty that enveloped him. It did not take long before he chanced upon a rope ladder. The rope stretched a bit when he pulled; as he looked up in the fog, he could not see its end. He began to climb the rope, keeping his head turned up. He stopped when he noticed that another rope was converging onto the one he was climbing. Soon he reached a node where several ropes joined from below. From there several other ropes diverged, some rising slowly, some steeply. To his left, in the distance, he saw a dim point of light.

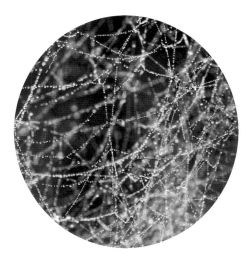

The voice resounded again. "Do not waste time," it said. "There are many nodes to climb and lights to see."

Which one is this? asked Galileo.

"You are climbing the exterior scaffolding of my palace," said the voice. "And now you are standing at the node of the first mechanism: what you see is the light it generates inside."

So be it, thought Galileo, and climbed one of the ropes that seemed to go straight up. When he reached the next node, he noticed another light far away, not far from where he had seen the first light. But he could see only one light. Am I seeing another light, or is it the same one?

"You are seeing the light of the second mechanism," said the voice.

"The light I see seems to be attached to countless filaments that fan out from the mist of absolute ignorance!" shouted Alturi.

Galileo could barely see the thin filaments.

"In my palace nothing is attached, but everything is in its proper place," said the voice from the heights.

What kind of palace is this? wondered Galileo aloud. I do not understand.

"Then I shall tell you this," said the voice: "The nodes and ropes you are climbing form a complex, they are its elementary mechanisms. They are connected in complicated ways, whereby they interact and turn each other on or off—all elements are off right now. Depending on their state, they generate information, and the position of each light specifies what information they generate—which states are ruled in and out. Now you understand?"

Even less, replied Galileo.

But from afar Alturi intervened: "I bet I do understand. The voice that's guiding us is strung as lean as a logician's thought, as if he might be made of glass. So here is my deduction: those filaments I see inside the palace are axes of giant space, one that lives not in three dimensions but in trillions. Therefore the lights we see are points that occupy special positions in this space."

"You are proof that arrogance can be wed to logic," said the voice. "There is indeed a space inside the scaffold—it is the space of qualia.

Each of the filaments you see is an axis in that space, and the axes are as many as the possible states of the complex. And I shall tell you one more thing: each filament is a measuring rod of glass that goes from zero to one."

"I see," exclaimed Alturi. "Then I know what this means: the coordinates that specify the position of each light inside this qualia space must be probabilities—the likelihood of past and future states of the complex, each one corresponding to a different axis. In that case, each light stands for a probability distribution—the repertoire of the past and future states of the complex that could have led to its present state, as specified by a particular mechanism."

"You are well ahead in the context of the bright," said the voice, sounding as distant as before. "The position of each light reflects a probability distribution, and each probability distribution over the past and future states of the complex is an irreducible concept."

"So a concept is just a way of grouping together the states of the complex," said Alturi. "States compatible with that concept are given high probability, the others are ruled out. I fully understand."

I don't, said Galileo.

"It's simple," said Alturi. "For example, from the perspective of the complex—what he calls his palace—some of its states are consistent with there being a chair in some of its rooms and are given high probability, but some are not, and their probability is zero. Therefore the concept 'there is a chair' is a light in the particular position in qualia space that corresponds to that distribution of probabilities."

"Just so," said the voice. "Then I shall ask you this: What does the complex know? Does it know where it is coming from, or where it might be going?"

Alturi did not think for long. "A complex is what it is—you said it. It is a set of elementary mechanisms—the nodes and ropes we are climbing—and now it finds itself in a certain state. What can it say? That it could have come to its present state from some of its possible past states, but not from others, as dictated by its own mechanisms. Where will it go? Where its mechanisms will allow. A complex knows what it can rule out, by virtue of how it's built: it knows its own concepts—which ones are true and which ones are false in its present state, as specified by the position of the lights—but is uncertain, or should I say it's wholly ignorant, about the rest."

Galileo reflected. A mechanism not only determined what would happen next; it also specified what might have happened before. Without mechanisms, none of the trillion states available to the complex was more or less likely than the others—all were one-trillionth likely. That was the mist of maximum ignorance. But with a mechanism engaged, some of the past states, those that could not have led to its present state, could be ruled out, while others became more likely. The same for future states. So each mechanism specified its own irreducible perspective on the probabilities of past and future states of the system—its own irreducible concept. That was why the light of the first mechanism was in its particular position, that of the second mechanism in a different position—a different probability distribution corresponding to a different concept. And so, thought Galileo, mechanisms reduce uncertainty and generate information, as he had learned from S. Emboldened, Galileo climbed farther along the ropes, from node to node: at every node, he thought, he could see the light it generated in the distance. What kinds of mechanisms are in action here? he asked.

"Galileo," reproached the voice, "how could you ask such a dim question? What mechanism focuses the diffuse gray of ignorance into a point? Your fortune was built upon a lens, and now you can't tell one when you see it in action?"

Galileo was stunned. Where are the lenses?

"You have walked all over them," the voice told him. "They are the elementary mechanisms, and they are set in the nodes where the ropes join, like precious stones set in their bezels. Each lens is kept at its position by the tension of the ropes that connect to it and pull in different directions. When its duller side faces the interior of the palace, the light is not focused sharply—the lens is off. But if it receives a sudden pull from many ropes pulling in the same direction, the lens turns its sharper side toward the interior—it turns on. In doing so it focuses the light more strongly, and at the same time it pulls the ropes that connect it to other lenses."

"So the turning of the lenses and pulling of the ropes—the scaffold of ropes and lenses that is the exterior of your palace—determines its new state; yet at the same time the lenses focus their lights somewhere inside your palace—the mechanisms specify their concepts and generate information. So causation and information are one and the same," said Alturi. "But how many lenses are there?"

"Ah," said the voice, "Galileo knows well that lenses are precious. There are only so many different kinds of lenses one can make. Perhaps I've made them all—my life spent as a lens grinder, safe in the obscurity of my art."

"Are there some lenses that are truly sharp?" asked Alturi. "The lights we have seen so far seem quite dim."

"What kind of guests are you, complaining of my palace, when you have only seen its entrance?" retorted the voice from the heights. "You have just seen the action of a few simple mechanisms, and you think that's all? Climb halfway between two nodes and look."

Carefully Galileo moved down the rope he had climbed and stopped halfway. At first he did not see. Then, squinting and moving his head, he saw something shine in the distance. It was hard to keep the image in his line of sight, but when the rope ceased to swing, he could discern another light, brighter than the previous ones.

"I know what this is, though I suppose Galileo may not know about interference," said Alturi's voice, ever more distant. "At each node one can see the light focused by the lens of that mechanism. Midway between two nodes, however, one can see something else: the light generated by the interference between the beams coming from the two lenses, which may be very bright if the interference is constructive. And one can see the light of interference only from the right perspective, one that conjoins the two elementary mechanisms halfway down the rope. Isn't it so?"

"Indeed," said the voice.

"What happens if the two lights don't interfere?" insisted Alturi.

"You disappoint me," said the voice. "Of course without constructive interference, no further light is generated. For how could it be otherwise? A concept only exists if it is irreducible to simpler constituents, or else what difference would it make? As is the case with everything: what is, is what cannot be reduced. Now climb, for only by striving can man be released."

Galileo climbed another rope, and then another one. Every time he reached a node, he could see the light it generated, sometimes bright, sometimes almost imperceptible. And at times he could see another light when he was halfway between two nodes. So Galileo went on, climbing rope after rope, all along the spider web of ropes—a mesh arranged as if along the surface of a giant sphere. And at its center, where the countless thin filaments originated, he saw light after light after light until, breathless and dizzy, he heard again Alturi's voice, now very close.

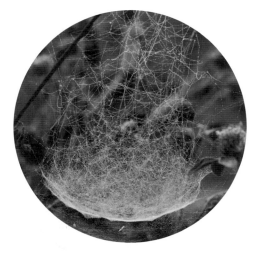

"Quite an expensive palace," laughed Alturi. "All this work, all this grinding, all this climbing, all this scaffolding of intertwined ropes pushing and pulling the lenses at their nodes, and all these lenses beaming in the dark, but all one sees from the perspective of a single node is a single light. Well, maybe it's not such a bad metaphor for something, no better and no worse than the German's mill. But isn't it cheaper to make metaphors of words, as poets do?"

Galileo was mumbling to himself. Are there lights generated by interference among more than two lenses?

"Of course," answered the voice. "The lights generated by interference between the beams of two lenses are quite a few, but there are many more ways for three beams to interfere, even more for four, five, or six beams. If their interference lights were all realized, you could never hope to count them."

"Why do we not see these other lights?" asked Alturi.

"You disappoint me again," said the voice. "You would need to share the perspective of a combination of many lenses, and that is hard."

"Be it as it may," replied Alturi, "when all is said and done, I still don't see your purpose—I cannot see how all this toiling with lights and perspectives could explain the thing that needs explaining—the quality of experience. Yes, every lens generates information, focuses diffuse light onto a point somewhere, and the brighter the light, the more the information it generates that cannot be reduced. Yes, interference between lenses may generate additional points, concepts that cannot be reduced to those produced by individual lenses. But in the end what? In the end, all you have are points of light, and a point has no quality. It is as N. was saying: you won't get consciousness out of matter. And neither will you get it out of light."

Meanwhile Galileo was looking around. What is this device? he asked. Attached to a swinging rope, he had noticed something that looked like a telescope.

"Ah," said the voice. "You have found the qualiascope, at last. For only with the qualiascope can you see all perspectives at once. Now don't be afraid, peer into the qualiascope and dive into the space of qualia, flying toward the light."

So Galileo looked and saw. At the center of the palace were lights in the billions, billions of shining lights, spread out over the giant space of qualia, each light occupying a different point, some brighter than others, and each of them was a concept, an irreducible concept generated by a mechanism—a single lens that focused light, or a combination of lenses that shone light by interference. He could not see the spider web of ropes, nor could he see the lenses; instead he saw the dots of light, collimated by the lenses. And so he thought: What mattered was not the ropes holding the mechanisms, what mattered was the filaments that outlined the space of qualia, one for each of the possible states of the system. Then, finally, he understood: It was not the ropes, not the lenses, not the filaments, not even the individual lights that he was meant to see; it was the celestial constellation—the diamond of a trillion faces, it was the shape unseen.

"Now you can see my palace, its vast, magnificent contours," said the lens grinder. "Behold its myriad lights. Though it may seem rising from the mist, truly it's a magic lantern, unfolding a field of light into the image of experience. Out of sheer ignorance a splendid shape here is created; out of gray, undifferentiated fog, a phantasmagoric picture is born, a picture of diffractions. Yet it is not in those diffractions that beauty lies, nor the meaning of my work. It is the palace itself that matters, the only one that has true meaning. It is how those diffractions relate to one another, the work that's done by every set of lenses, in which direction, and how bright, they do project the light, and how they interfere, generating yet more lights. It is the shape that matters—a crystal kaleidoscopic, conjured by a chrysalis of mechanisms; a crystal of beauty phenomenal, begotten by the light that shines through a trillion lenses. It is the quale—the shape that is the quality of experience.

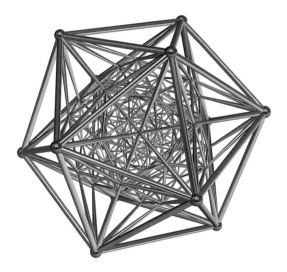

"How can one understand such verbiage?" shouted Alturi.

I think I see it, said Galileo. Each experience is a shape, and like a shape, it is what it is: a pyramid is not a cube, a sphere is not a tetrahedron. A shape built of concepts—points of light that shine more or less brightly, a single, extraordinary shape, as long as it is one and does not shatter. A shape that specifies, uniquely, the quality of experience. And still it's hard to see. Because the shapes we know are those of bodies, and rocks, and things, the ones we see and touch, always from the outside. We never think of darkness being a shape, not flat and simple, but richer than a diamond; of thoughts and wishes being yet other shapes, fantastic shapes; that my consciousness, my present feeling is a shape, that I'm myself a shape, a shape extraordinary, the only shape that matters, but is not made of matter.

Galileo was short of breath. He paused to recollect his thoughts, and asked: If you can tell me, lens grinder, say how large your palace measures.

"Like an enchanted castle, Galileo, mine floats in the air, and it has no weight. What matters is its shape, and with shapes this is what matters: how much it takes to capture all its points of light, to find all corners of my house."

Like an enchanted castle, thought Galileo. But tell me, he asked again. Seen with the qualiascope, the shape of the palace seems made in turn of smaller shapes—lights that gather together in space—and though they may themselves be clustered on a wider scale, each of them is different.

"Yes," replied the lens grinder. "Every palace has wings, and those of mine are sight, and sound, and smell, and touch and taste and many more, and within sight color, texture, and motion, and within color the sub-subshape of red, a splinter of the palace that can't be further divided into parts. And some subshapes are similar, like red and blue, though they are not identical, and differ more from the subshape spawned by the lute's third string. The castle is always one, but many are its corners, and many its secrets."

At that Alturi shouted again—his voice was shrill. "Which is the splinter that is red?" asked Alturi. "I want to see the shape of red."

"No more will you have red without its palace, than you can have a hole without its surface, a figure that has lost its ground, a grin without the cat," said the lens grinder. "And now behold this sight."

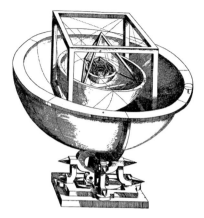

At which a million lenses turned, forming new points of light and new diffractions. Before long a new shape had formed, a new experience, a novel quale.

"I kept my castle still so you could climb it, quiet, its lenses off—none of its concepts verified—a lasting state where consciousness is full without having content—a castle meditating. But now you see: the mechanisms push and pull—determine a new state—lenses turn on or off, and so changes the light they shine, to specify the countless points—the countless concepts: one beating of the heart, and my cathedral changes, another one is formed, the shape transmogrifies. And so it goes, the castle has within itself the power to restring its mechanisms, retune its lenses, and generate another castle from its ashes. So from its eye I watch my palace change, the constellation, the magic lantern, the loom enchanted, that weaves its dream all of its own. As long as I may live, I want my palace tall, so that my thoughts can breathe unfettered within. And when the palace crumbles, I'll plunge with it toward the void."

The master grinder vanished, and Galileo was left alone, surrounded by the night. As when emerging from a dream one tries to catch a thought that flies away, so his own mind soared with his hand, to carry back a vision from within. And so he did, and wrote not to forget, and this is what he wrote:

The many mechanisms of a complex, in various combinations, specify repertoires of states they can distinguish within the complex, above and beyond what their parts can do: each repertoire is integrated information—each an irreducible concept. Together, they form a shape in qualia space. This is the quality of experience, and Q is its symbol.

NOTES

After Leibniz, Spinoza, as if this were the right sequence! The voice from up high is Spinoza's beyond doubt, since he worked as a lens grinder much of his life. Perhaps the affinity for lenses he shared with Galileo is the reason Spinoza is treated so reverently in this chapter, or perhaps it is because in the *Ethics* he said: "In proportion as a body is more capable than others of doing many things at once, or being acted on in many ways at once, so its mind is more capable than others of perceiving many things at once" (Scholium of Part II, 13). Under a very generous interpretation, this proposition could be construed as suggesting that the ability to distinguish among many possible states is related to the presence of consciousness, very much to the author's liking. Or maybe it is Spinoza's fixation that everything could be demonstrated "geometrically," as if he were the Euclid of ethics! Whatever the reasons, it is hard to accept that the epiphany of the phenomenon, the revelation that experience is the shape of the quale, a shape made of integrated information, the quantity of quality, should be entrusted to Spinoza! Such an aloof, elitist character, a fatalist and a denier of free will! Why not Leibniz, who had made very much the same point in the previous chapter, who was a true optimist, a believer in freedom, and who excelled in so many fields? Worse, Spinoza was a totalitarian ascetic, believing there is only one substance, Nature = God, while Leibniz was a true democrat, a true pluralist, who loved all aspects of life (or almost all), and who indeed proved that there is an infinite plurality of substances, the monads, and that monads are everywhere in nature—humans, animals, plants, and even rocks (though it's good that some monads are more dominant than others). Or take Spinoza's theatrical, self-centered empty ending, "I'll plunge with it toward the void," against Leibniz's modest, warm, and constructive "the space where feeling is defined by mechanism." Once again, Leibniz is shortchanged, just as when he was accused of taking secret inspiration from Spinoza, while instead he took it from

Bruno. Worse still, Spinoza had nothing to say about what determines the internal structure of experience. Leibniz instead championed a "relational" theory that extended even to space and suggested that relational properties had to be based on intrinsic properties of his monads. True, he too had nothing to say about the relational properties that were intrinsic, but then who did? At any rate, the choice of Spinoza may explain why, despite its ponderous, self-important style, the chapter fails to convey a straightforward message that everybody can understand: each experience is a shape made of irreducible concepts, which are probability distributions of past and future states of the complex, specified over the space of qualia by the causal mechanisms that make up the complex in its present state. Of course, the scaffold/spider web of nodes and ropes holding the lenses represents the set of causal mechanisms—such as neurons and connections in the brain. The constellation of lights they generate—the palace of light—represents the shape of experience in qualia space. Each light is an irreducible concept —a point of integrated information—that says which past and future states are compatible with the mechanism's present state, and which are not. But why resort to such an elaborate, unintelligible metaphor for such a simple idea?

The 600-cell (Robert Webb, www.software3d.com/Stella.php) is the convex regular 4-polytope, the 4-dimensional analog of the Platonic solid icosahedron. The vertices may represent possible points in qualia space—one for each irreducible combination of mechanisms or lenses—where each point stands for a different probability distribution. Here it may be meant to illustrate the shape of a very simple quale, the geometry of interesting qualia being incomparably more complex. The "Celestial Rose" image is from Dante's *Paradise* illustrated by Doré. *Haze* is courtesy of Nancy Lobaugh. Kepler's Platonic solid model of the solar system is from his *Mysterium cosmographicum* (1596). The spider's web was identified as the product of *Arachnis qualiatextor,* a rare species found near Verona.

21

THE GARDEN OF QUALIA

*In which is said that the universe is mostly dark,
but the largest stars are closer than one thinks,
if they are looked at with the proper instrument*

Consciousness came back to him as darkness lingered outside. Feeling
the earth around him, Galileo opened his eyes and saw the stars. At
once he was aware that he was not alone: next to him stood a woman,
her face was like the moon.

And like the moon, her face, lit by a quivering lantern, was full of
craters, vestiges of eruptions, forgotten long ago. Thus thought Galileo, but who was she, a woman wrinkled, like an old hickory tree, and
shorter than a broomstick? Why was she brandishing a looking glass?

"Watch, Galileo, watch," she said. "This nebula is beautiful, just as
I never was. I counted many, but for this pretty one, I had to sweep
a lot of dust."

The nebula was beautiful indeed, thought Galileo. Hard to think that the old woman, who must have spent her life bent on her knees, washing and scrubbing, would know about the sky.

"The sky is full of nebulae, and each of them contains millions of stars, and many of them are larger than the sun."

How come you know these things? asked Galileo.

"I had to polish many lenses in my youth," she said, looking downward. "At times I looked through them and wrote down what I saw."

Who taught you to look and watch the stars? insisted Galileo.

"My mother never wished me to," answered the woman. "Like Cinderella, I was to be in the house for life. But Father saved me, he taught me to strive always, and to better myself, when Mother slept.

Then, when I was still growing, an illness struck me and took away my looks, and I could grow no more. No man would ever want me, except for one, who took me under his wing. My brother took me, first I was his maid, but then he saw the spark that was in me and stirred it, and fanned it, and tended it. So I resolved to grow, to grow by learning, and learned some words and many numbers. And then one day, after he had taught me how to sing, my brother took me out, for all to see, and all to hear, and they did not dislike it."

Your story is strange and pitiful, said Galileo, but tell me, who taught you about the stars?

"My brother. He wished to look into the sky, deeper and deeper, and he began to make his lenses, and he was good at it, and I learned it, too. He soon forgot about the music, even about the food, and I fed him, food, lenses, and paper, I fed him all night long. So he discovered many stars, farther and farther away, there was no end in sight. He said to me that one should go look for the stars, what else were we here for? So when he was asleep, I went and looked myself."

I too have polished many lenses, said Galileo, and looked through them, and wrote down many things, and now I know that with my lenses I only saw the stars that are closest to us. But I have come to this garden to look through a different kind of lens. A master grinder gave me a qualiascope.

"Ah," said the old woman. "You want to peer inside all living things and know their secret shapes. Then come with me and walk this ancient garden, but do walk slowly, as I am weak and there is much to see."

I will, said Galileo, but first I'll point the scope toward the sky, to see what shape the stars and nebulae may hold. And so he did. But no, he felt, the stars had let him down, even his Milky Way. Up close, he thought, each star would be more than immense, magnificently full of mass. But when it came under the qualiascope, each star turned into a flimsy thing, a dim handful of dust. And when he roved the scope from side to side through the vast universe, he found it mostly empty; even its largest bodies, the constellations that had been revered, the galaxies that had been discovered, infinite spaces beyond the garden's hedge, lost all their gravity, dissolved into gray powder, and had no shine in them.

The woman raised the lantern, so he could watch a moth, diaphanous, small and silly, as it was gyring toward the flame. Galileo raised the qualiascope and looked in the direction of the moth. A moth, a moth was wider than a star! A firefly that would carry a sun! Like a diamond with a trillion faces, like a spider web of light, like a flame enveloping a temple, the quale inside the moth was sparkling brighter than the firmament. One it was, and it was not dust.

"You'll look in vain at rocks and rivers, clouds and mountains," said the old woman. "The highest peak is small when you compare it to the tiny moth."

So Galileo pointed the scope around him and looked for qualia. Under the scope, the garden was a thin gray vapor, a vapor that didn't lift or breathe. The sun was dawning, but the sky stayed empty. He looked at a majestic tree with his left eye, then closed it and peered through the qualiascope with his right eye. The tree dissolved into a thin dotted drawing, traced with a sharp gray pen into the scarce morning light.

He looked around and turned a dial to enlarge the scope's view. Out of the earth's dark vapor rose other diamonds bright, or brighter, than the moth's. But brighter still, much brighter, and grander on a giant scale, was the owl's own sun.

"It is a comet like I never saw, that's carried by the poor owl's head, flaming through the morning sky, more intricate inside than a lace of laces!" cried the woman. "My comet is a trifle held against the owl, and nothing when you view it through your scope." The woman had turned off the lantern and now was pointing to the woods. There, steeped within the mist, were other comets and other constellations, a galaxy if you caught it all. Rising from sleep, the animals were turning on their lights.

Inside the fires are souls, remembered Galileo, each one clothed of flames of which it burns. Dancing, quivering flames, their shape changing without end. Cathedrals of fire, built and rebuilt at every moment. Each shining from the inside, of that light that can see, not light that needs an eye in order to be seen.

    The woman took him by the hand. They were by now outside the garden. But where? A convent, a hospice, or a graveyard? There lay Copernicus, his head bearing the faintest candle. Next stood Poussin, his quale trembling all aflame, and the blind painter, shining of inner light, vaster than his masterpiece. And there was M., his old friend, the measures of his quale dwarfing all the primes. Then he saw Ishma and El, their split two-horned flame dancing like Ulysses and Diomedes; he saw Teresa, her flame split and unsplit by icy gusts of wind; the witch girl's flame, seized by such a frenzy that it had lost all shape. And then at last Galileo saw the philosopher—he was no more than embers tamed by sleep. But then his fire revived and surged into the heights, like a burning cathedral, and so he knew that his dream was—he was his dream—his dream heavier with being than all the stone and glass of Chartres.

So Galileo understood and turned the scope toward the old wizened woman. A slender spire of dust rose from the dark floor, and over it, behind craters and wrinkles, shone Caroline's constellation, shaped with a million symmetries—the soul's own naked beauty outshone the garments of the flesh.

NOTES

The old woman is Caroline Herschel (1750–1848, portrait by M. F. Tielemans), the sister of astronomer William. The story she tells is true: at ten she had typhus that left her disfigured and stunted. After her father's death her mother kept her working in the kitchen, until she moved to England, where her brother worked as a choirmaster. William became interested in astronomy to pass time at night, and she followed suit. Caroline helped William with his large new telescope and once was caught on a hook, "leaving behind nearly two ounces of flesh," when they finally lifted her off. She discovered several comets, including the periodic comet 35P/Herschel-Rigollet (the one in the picture is Comet Hale-Bopp, taken by Philipp Salzgeber), discovered galaxies such as the dusty NGC 253 ("Astronomy Picture of the Day," NASA, November 21, 2009, courtesy of SSRO), and compiled a catalog of stars, although she never learned the multiplication tables. The fireflies are courtesy of Kristian Cvecek. The owl in full moon is by Kirk Woellert/NSF. Caroline was the first woman member of the Royal Astronomical Society and received a pension from King George III, the first time a woman was recognized for a scientific position. Galileo, as usual, is mindful of Dante, and when he says "Inside the fires are souls, each one clothed of flames of which it burns," he is referring to *Inferno,* Canto XXVI, *"Dentro dai fuochi son li spirti; catun si fascia di quel ch'elli è inceso."* There Ulysses and Diomedes did in fact share a split flame; the illustration is by William Blake (at the National Gallery of Victoria, Melbourne). The constellations at the end are taken from Pesello's cupola of the Sagrestia Vecchia in Florence.

PART III

{IMPLICATIONS}

A Universe of Consciousness

22

INTRODUCTION

Sparks and Flames

Like a tree with a million branches, the great cathedral sent up its spires. Built over as many centuries, some had crumbled; some, imperiled, were suspended over the void. Others had risen of narrow, contorted stems to attain sublime heights. No architect could have been so daring and so blind to consequences. No constructor could have wasted so much wealth and life. Yet there it stood, in a rain of stones plummeting from its lofts, threatened by winds and quakes.

But all around was dark and empty. Then a light beckoned, and like a hopeless moth Galileo advanced toward its source.

Shaped like a diamond with a trillion faces, alive like a raging fire, a light containing all colors and all shapes, all words and sounds was shining before him.

A woman veiled, her countenance half hidden, was tending to the flame. "In darkness infinite and eternal, a flame is all we have; a flame is all we are. It is the only precious thing. I am one of many, scattered in this immense and crumbling palace. If the flame perishes, it will be my end. Majestic wings collapsed to ruins, in times past storms have brought destruction, cities of light were blown away, but children rushed, carrying the sparks, and new fires were lighted in the void."

So Galileo saw them, dimly in the dark distance, like a luminous rosary, like a constellation gyring into the heights. Connected by corridors and chambers, each sacristy led to another, and then to another, and at each fire a Vestal knelt.

"Every flame is sacred, every spark," said the old man. Like a sage, like a prophet bearded with a mane remembering the inflections of his words, he sat in candlelight writing the story of fire, how it was sparked by lightning but gave no light, for million years it gave almost no light, and then it grew, and begot light shining bright, and brighter. It shone brightest in man, he said, but not for this was it any safer—as it could burn itself to death.

NOTES

For some reason, the name of the third guide, Charles Darwin, is never mentioned, and he is merely referred to as the bearded old man (as in the photograph by Julia Margaret Cameron). Altogether, it is curious that, of all people, the guide in this third part should be Darwin, and that the author has him proffer free advice, of all things, on what is sacred. Notoriously, Darwin's theory has been accused of undermining any sense of the sacred, indeed, of emptying life of genuine moral meaning. True, some philosophers, such as Dan Dennett, have deployed the joint powers of logic, evidence, and persuasion to argue that Darwinism does not destroy meaning but puts it on a new, better foundation. And scientists such as Gerald Edelman have suggested that the brain itself changes and grows according to Darwinian principles, so we should cherish the variety of nature and the quirks of the individual. But in this third part of the book, which deals ostensibly with the human implications of viewing consciousness as integrated information, would it not have been better to resort, for once, to a guide of more spiritual disposition? Crick, Turing, and Darwin, one wonders: Isn't this trio perhaps a bit one-sided? And, as for variety, also a bit too English? *Magdalene with a Night Light* by Georges de La Tour is at the Louvre, Paris.

23

NIGHTFALL I: DEATH

*In which is said that, if consciousness is integrated information,
it dissolves with death*

The fire extinguished. And the flame extinguished inside the fiery head, its brief flare consummated. Ash and ambers left, poor dust to bear witness to the soul. Ash rose in clouds—ash that could be touched, unlike his soul. Ash that could be held in an outstretched fist, as his will could not. Ash that was light and airy, but with no breath in it. And it was surging high, above the buildings, above the trees, engulfing the indifferent city and darkening the sun.

Bruno was burned. But Galileo still was, his eye seeing, his heart feeling, his mind begetting thoughts bleak of the intercourse with death. Bruno had been a fool. Why end it, if nothing was in sight? Why play the martyr, if he was right? If it was immortality he craved, it was the dubious immortality of words—the immortality of philosophers. To a scientist, immortal was the truth, and truth needed his work, all of his work, till the end of life. So Bruno had been wrong, not Galileo. Bruno had been proud in death, but Galileo was wise. Or so he thought, because his heart was not in it.

They had taken Bruno in the morning, the spirit boiling inside his flimsy frame, bubbling with images and visions that overflowed into cascades of burning words. So they had pilloried his darting tongue tight, and his thoughts were swelling against the stiff walls of his head. His consciousness bursting in unrelieved pressure, he thought he might break free and scatter into the universe. But when they had arsoned his body, he was surprised he did not feel a thing.

There was nothing to fear, they said, as no rational being can fear a thing it will not feel.

Not seeing that this is what he feared—no sight, no sound,
No touch or taste or smell, nothing to think with,
Nothing to love or link with.

Dissolution was irreversible—perhaps irreversibility was a law of the world—not a law but a sentence, a guilty sentence that punished all. Would it be better to step onto the stake of his own will, as Bruno had done? Instead he had delayed dissolving into nothingness for a scant reprieve—so that something was, rather than nothing, for just an instant longer. So that a light would flicker briefly in black eternity. Not pride, not principle, death was the greatest loss.

Whose loss? Of her who died or him who mourned her? His own dying daughter had consoled him thus: his were the pain and memory to hug a shadow as if it were made of flesh. And yet his light endured; but hers? Beheaded was her sunrise, usurped by a night that knew no dawn. A whole sea of black ink had drowned her words—there was no more to read. No late clock would annoy her now. The peaches in the convent garden ripened for no one. How quick her hand undraped the window—that morning was the last she let the light shine in. That candle now would burn no more. She had been born the year of Bruno's death.

His own death, too, when it would come, would be the ultimate. Worse still if it had come vengeful, wretched with impotence, impotent toward the future. He saw them coming, taking her and spoiling her and smothering her and severing her from his memory. It was consummated.

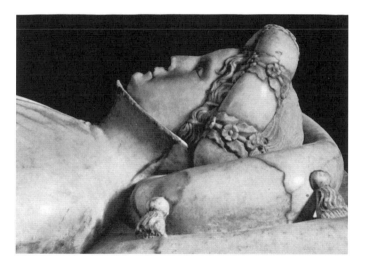

So Bruno's image had led him astray, through a wide gate of bronze, to an immense dark hall. The floor sloped gently, it took no effort to follow it downward. Straining to see, he made out a faint shimmering of stone, reverberating over endless slabs that vanished in the distance. His hand shivered touching the smooth cold marble. She lay simply and quietly, her eyes closed, her features frozen. That such goodness should perish, that anything should and would, that was the law, the most immutable among the laws.

And Bruno carried him further down, where endless gravestones disappeared in unbound dim rows, each gravestone with just a name inscribed. And then there was a dark wall, a wall that sank into the ground forever. His hand was over it, touching name after name, one after another, a never-ending list. Then, at great distance, the wall dissolved into a night of fog. In the fog there were no names—it was the night and fog of those whose names were lost.

All his life he had played with numbers, with numbers large, numbers astronomical. But now he struggled to swallow, as if that number he could not take in—the number of the dead. No, that number was not a single number, but it was made up of ones, countless ones—each one a universe of consciousness. How could he embrace all those ones within the tenuous circle of his empathy—a circle where he sole stood, suffused of direct light, his dearest ones of light reflected and diffracted, and darkness hid the face of all the others?

And yet each of those others had once been animated, by inner light as bright or brighter, by an intense thrill of being, as vivid and as violent as his own, throbbing in the private of one's consciousness. The simultaneous strands of sorrow emanating from a million selves, scattered sick around him, how could he ever follow them in life? *He had not thought death had undone so many.* Subjectivity multiplied, its inner import exceeding all his insight, the extent of his compassion exploded by the enormity of their dissolution. And how could such a loss be measured? How should it? A number for everyone, adding down to zero.

So that was what death was. The extinction not of life, not of movement, but the extinction of that inner light, which no external blaze could resurrect. Because death was the loss of consciousness forever. That was what death was—when information was torn to pieces in the brain—disintegration of the body was disintegration of the soul.

And death was not what it used to be—the heart stopped beating, and breathing ceased, the vital systems of the body all failed together. The year he had been born, the year Michelangelo had died, Vesalius had carved a nobleman in front of a great crowd. The man was dead, but when the chest was opened, the heart was beating still: the brain was dead, but not the heart, and Vesalius had fled. What was death now, now that the body could be kept alive long after the brain was done with? A woman's heart continued beating for two months after her brain was cold, and she was delivered of a live child. Thus doctors had learned to probe brain death with care. The dead must stop breathing, pains produce no response in limbs, or in the face and tongue; the eyes cannot be made to move, and pupils don't respond to light. Yet for a few hours the spine may spasm and twitch: like Lazarus, the dead sits up, stretches his arms, and fools the living.

Yet to believe in death was hard. They all had thought there was the body, and then there was the soul. And consciousness could not but side with soul. Oh yes, his body may perish for eternity—burdened with age and broken in its joints—damaged goods already, it had betrayed his soul. But the soul, how could it rot?

A consolation unworthy of a thinking man. For when the earthly Galileo would die, which Galileo should have survived in spirit? The shy and modest youth, or the man proclaiming his discoveries? The arrogant scholar of the world, or the bitter recluse of his later years? The one whose eyes had seen Jupiter's moons, or the one whose failing eyes could read his daughter's lips no more? The frightened one, bearing witness to his body's death?

He thought back at what he had seen. Experience vanished with a mere blow to the head: one day Copernicus's soul had dissolved together with his brain. Every day his own soul ebbed and flowed with sleep and wake in his own brain. He remembered the old paintress, not just blind but with no notion of what seeing meant: a stroke to the back of her brain had left her soul an amputee.

So consciousness was born with the brain, flourished when the brain grew its green connections, pruned and refined them to build the scaffold through which the shape of qualia bloomed, and then had aged with it, the green canopy wilting, and soon enough the soul would die when the brain went dry. Yes, consciousness did not reduce to matter—Φ was the most intrinsically irreducible thing there is, the only thing that's really real. But consciousness did rely on matter, and if the brain was undercut, the soul too would collapse.

So why had he believed that the soul could escape death? Why had Dante believed it, conjuring clouds of spirits to yawn incurious sameness where all was stale eternally? Why Newton—born the year of his death, to enchain the universe to laws of his devising? Not faith, but lack of imagination. Unable to imagine how the perfection of the body could be built out of random variation and selection, they fancied a dexterous demiurge. Unable to imagine how the lute's ethereal sound could spring forth from the brain's dull matter, they fancied a transcendent soul, a soul set there to listen, a soul that would spring free with death. A soul that played the thin strings of the brain as a musician played those of the lute, and when they broke, he sang.

Perhaps as long as no explanation could be imagined, as long as science was unequal to the task, the soul could still survive in its secret sanctuary. But if consciousness too yielded to reason, then, like a fog swept by a cold wind, the mystery would dissolve, and death was certain. Perhaps. Because he had been wrong before. Tides proved the earth revolves around the sun, he thought then, but he was wrong. Perhaps the soul divorced the body when it was betrayed. And perhaps the sun revolved around the earth.

NOTES

"Nightfall I" takes us back to the heavy baroque atmosphere of Part I (a dark baroque without any of its resplendent gold). Giordano Bruno, naked, his tongue in a gag, was hung upside down and burned at the stake in 1600 in Rome, Campo dei Fiori, where his statue stands (by Ettore Ferrari, photograph by Jastrow). Bruno held Copernican views and thought that soul pervades the entire universe. The inquisitor, Cardinal Robert Bellarmine, had demanded a full recantation of his heretic doctrines, but Bruno refused. The quote is from Philip Larkin's "Aubade." The head of Ilaria del Carretto is from her sarcophagus in Lucca, sculpted by Jacopo della Quercia. Galileo's daughter Suor Maria Celeste was born the year of Bruno's death (1600), Newton the year of Galileo's death (1642), and Galileo the year of Michelangelo's death (1564). The wall is from the Vietnam Veterans Memorial in Washington, D.C. (photographed by Glyn Lowe), the photogram from Resnais's *Night and Fog* (*Nuit et brouillard*), and the bas relief is *The Massacre of the Innocents* by Giovanni Pisano from the Church of Sant'Andrea in Pistoia. Brain-dead patients can occasionally sit up due to a spinal reflex (Lazarus's sign). *The Resurrection of Lazarus* (detail) by Caravaggio is at the Museo Regionale, Messina. The *Rondanini Pietà* (detail) is at the Castello Sforzesco in Milan. Michelangelo worked on it until the last days of his life.

24

NIGHTFALL II: DEMENTIA

In which is said that consciousness disintegrates with dementia

Like a broken bough drifting in the stream, life had passed by the brief shores of memory, till evening had come.

Night's shades had fallen inside unfamiliar walls. A friar made way, stooping in the long corridor. He led Galileo into a dim room that stretched in the distance and seemed bare. Only a crucifix hung on the washed wall. On the floor was damp straw, the acrid smell of a stable. In a far, dark corner was a mattress also of straw; over the mattress a wooden cage made of rough boards was hanging, an iron chain suspending it from somewhere in the ceiling. Galileo drew closer. The boards formed a loose pillory: out of an opening dangled a wasted hand; from the two lower holes hung two bare feet. There was another hole above the cage. Straining his eyes, he made out a head, covered by sallow, beardlike hair: there inside the cage hung a small scarecrow of a man. Who had withered inside that skull, as hollow as the cage, and peered emptily out of veiled eyes?

The skull betrayed a face that stirred memories, and when the recognition came, Galileo shivered: it was the head of Cardinal Bellarmine, the head of the Holy Office—it was the head who had pronounced Copernicus an enemy of the Church, the head who had condemned Bruno to burn. Galileo felt anger and then pity.

The friar, his name was Finali, explained that the Cardinal had made a vow of poverty: this was the room he wished for.

Had he wished for pain and misery, too? asked Galileo. Why was he in the cage, hanging like a haggard bird?

Finali lowered his head: "The doctors say he should not hurt himself, and so he hangs, as our Savior once hung from the cross."

The Cardinal seemed half asleep and did not move or speak. Galileo remembered him perfectly: a castle of knowledge and a stronghold of the Church. When the Cardinal spoke, he spoke as if weighing each and every word. Words had consequences, especially his words, and they required reflection.

Perhaps the Cardinal was as reverent toward the Truth as Galileo himself had been. *In veritatis amore,* he used to say. Only the Cardinal thought the Truth was in the words—Galileo thought it was in the facts. And the Cardinal thought every Christian had it—Galileo thought he had found it himself.

"His Eminence was the most learned man in the Church," interjected Finali. "He knew the Bible by heart, he knew all of Saint Thomas, and he knew his science, too."

I know, said Galileo, but he also thought he knew what he could not know, as when he knew Copernicus could not be right, because if he was right, he said, we would fly off the earth like ants crawling around a balloon. He knew but he was wrong.

"Too much learning can become a curse," replied Finali. "The Cardinal used to lock himself in his room to write for days on end. But Jesuits are not allowed to work more than two hours without breaks, so all his life, every third hour, he flipped his quill in the air, caught it, and went on writing. He wrote as eloquently as a theologian and as simply as a child: *The Art of Dying Well* was his noblest book, the one he knew would please his God the most. But now it seems to be God's wish that he should die artless and unwell, worse than a caged brute, a poor relic abandoned by a precious mind."

Perhaps he burned his brain by too much study, said Galileo.

"Perhaps all human knowledge begets pride, and a proud mind must be broken, broken on the wheel of fate," said the friar. And then he added: "I found this among his later writings: I've underlined some passages."

In the dim light of a candle, Galileo could read just a few sentences: "You were not made to live like brutes, but to pursue the virtue of knowledge," and then he read: "If God held all Truth in his right hand, and in his left the steady drive for Truth, but with the clause, that I would always and forever err, and told me: Choose! I should humbly fall upon his left and say: Father, give! Pure truth is surely for Thee alone." And then he read: "Because only the man who aspires and strives can be redeemed." And finally he read: "To strive, to seek, to find, and not to yield."

The friar looked Galileo in the eyes: "I am not sure what it might mean, but it makes me afraid, afraid that in the end he lost his faith. Was this the peak of wisdom or the start of his downfall? This is the last sentence I could make out: *Cedite opes, abite gloriae, ite litterae, ite: Broken is the thread of thought, knowledge makes me sick in the stomach.'* What should one make of it?"

It means that thought and the stomach, truth and digestion, do not mingle well, said Galileo coldly. What did he write after this?

"He went on writing for seven months," said Finali. "He wrote and wrote, flipping his quill in the air, but he could not catch it anymore and spent his hours on the floor searching for it. Yet by the end he had filled seven large tomes, writing so small he could not read the script himself—I certainly could not, but he just went on writing: inside those tomes, I am sure of it, he poured the sum of all his knowledge."

Show it to me, friar, said Galileo, and when Finali handed him the first tome, he opened it, and could barely see what was in it. So from his pocket he took his magnifying lens and saw some words he could read, strewn haphazardly on the page, between them a shaky line that he could not decipher. So he read *Cenodoxus* and *emunction* and *drintling* and *philautia* and *incurvatio* and *ganch* and *pleroma* and *stover;* and then *cenodoxia* and *entelechy* and *jactantia* and *vapulate* and *voraginous* and *vespertilian* and *verbigerate* . . .

Give me the last tome, said Galileo, and when he opened it, the writing was even thinner, the shaky line between the words was gone, and the words were short, far too short, and when he looked closer he saw what it was: it was the letters of the alphabet, the Greek alphabet, repeated page after page: for pages it read O O O O O O O . . . , then it read Π Π Π Π Π Π Π . . . , then P P P P P P P . . . , then Σ Σ Σ Σ Σ Σ Σ . . . , then T T T T T T T . . . , then Y Y Y Y Y Y Y . . . , and then he could not read any further.

What do you say of this gibberish, friar? asked Galileo. Worse than the notebook of a schoolboy. Once he was learned, maybe he earned acquaintance with the truth, but now he knows nothing at all.

The friar hesitated, hurt by what he had just read, and said: "So it is true in the end, his mind must have been broken already: and it was broken because of pride—the supreme sin among the seven sins, the insidious sin that tries the soul, the sole sin that tempts the saints—it was because he lost his faith in God."

No, Friar, it was because he lost his glue, said Galileo, the glue that held his brain together. Say rather, Friar, how the Cardinal has become like this.

The friar bowed: "At first it was his memory: I found him more than once lost in the chapel, peering behind the altar with frightened eyes, not knowing his way out. And then the stern reins of his judgment slackened. But a cardinal has nowhere to hide, so one day when he was serving Mass, a rich merchant came to take communion, a man we did not like because he was one of those who declare the greatness of their faith by the size of their house.

"The man was kneeling, his eyes half-shut but aimed toward the heights, his hands were closed in prayer. His Eminence was dressed in his best robes (although his miter was askew). But then His Eminence stopped, took the host, and instead of saying the customary words, he asked: 'Do you truly believe this is the body of the Lord?'

"The merchant opened his eyes in disbelief, then closed them again, and with a baffled voice, but a pious face, he said: 'I do.'

"At which His Eminence slapped him in the face from left to right with sudden glee. I was nearby—perhaps except for me nobody had noticed—and I could see the merchant reckoning within himself, whether his reputation was more at risk if he did not react or if he made a scene. But as he did not move, His Eminence took the host and asked him: 'Then tell me this at least, do you truly believe your soul will survive death?'

"The merchant out of instinct moved his head to the side and did not dare to answer. His Eminence raised his voice, and all could hear: 'Do you truly believe your soul will survive death?' he shouted, and the great names of the Church knew not which way to look.

"The merchant watched around furtively, and in a hushed tone he said, 'I do, I pray you, I do, just do not become upset.' But he had no sooner finished his answer than His Eminence slapped him again: 'Vainglorious hypocrite, don't you know your soul looks like a turkey's? It pecks its way close to the ground, shaped like a peg, it pecks and pegs, peck by peck it pegs away. Soul of a pecker! A pecker blinkered by its engorged snoods, your eyes can only see the ground to peck, and nothing more. You selfish gobbler! I see them well, I see the waddles and caruncles on your soul!'"

Finali paused, as if afraid that he had said too much. But then he continued: "That was his last Mass. After that his memory lost width and depth, and I knew it, and he knew it, but then his reason suffered more, soon all his faculties faded away, and slowly he lost his spark, spent many hours doing nothing, not out of sloth, no: because he knew not what, or how. And after that he forgot how to dress, and how to walk, and finally he forgot how to eat: he munched water and swallowed food whole."

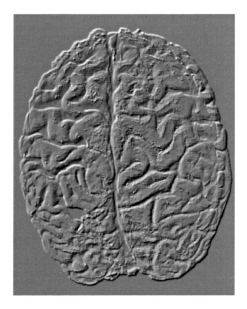

Does he know where he is now? Does he recognize us? asked Galileo.

The man in the cage showed no reaction.

So has his mind fizzled by now? Galileo pressed on. Where did it go? To chase after his soul in Heaven?

The friar was hurt. "His Eminence drifts off to another place: at times it seems he is half in Paradise. But though his earthly words or deeds cannot defend his mind before a haughty judge, his soul deserves respect."

Can you hear me, Cardinal? Do you recognize me? Galileo spoke into the Cardinal's ear and asked firmly, with his full boisterous voice. The man's mouth barely moved and he groaned. The eyes closed but the lips formed a few mangled words:

... sic ... non ... sic ... non ... sic ... non ... sic

Yes, Finali, he hears! exclaimed Galileo. And he knows who I am!

"Then these are his seven last words," said the friar. "Like our Savior on the cross." And then he said: "But this is only what makes it outside: God knows how much is left within."

There was silence. Then the Cardinal mumbled something again. This time Galileo listened closely and thought he understood this:

. . . sic . . . aut . . . non . . . sic . . . aut . . . non . . . sic . . .

This way or not this way, repeated Galileo. So that's what the Cardinal had to say. His ultimate wisdom is the wisdom of a diode.

"What is with this diode?" asked Finali. "Perhaps it means His Eminence will see God?"

I do not know, said Galileo. This way or not this way. How would anyone know?

But Galileo's mind was elsewhere. Again he saw Salerno eroding the brain, he saw the vast, magnificent complex of the mind crack under the sick teeth of age, he saw it crumble into a cemetery of diodes. The full measure of Φ may be hard to fathom, but like the Cardinal's spent face, its number must wither, its repertoire disintegrate—like a chandelier smashing to the ground—the higher it shone once, the more the crystal splinters on the dirty ground; he thought that when the brain is shattered, the mind shivers into dim fragments, that when the cortex shrinks, consciousness is bound to wilt, and when the skull is emptied, the soul too bursts, like a balloon into the void chest of the night.

NOTES

There is no hard evidence that Robert Bellarmine, Galileo's inquisitor, died of dementia—perhaps the Cardinal is meant to stand for

Everyman. This may also explain why several sentences attributed to the Cardinal are a strange medley of Ulysses and Faust, of Dante, Lessing, Goethe, and Tennyson. Bellarmine was a very learned man and did in fact write *The Art of Dying Well* and *The Seven Words on the Cross. Cenodoxus,* one of the words written by the Cardinal, a play in Latin (1602) by the Jesuit Jacob Bidermann, tells the story of a man of immense learning who has a reputation for charity and saintliness. But he is struck by disease and finally death, and his body in state refuses to be buried, raises itself up three times, and finally cries out that he is being accused, found guilty, and condemned to Hell, because of the deadliest of sins: pride and vainglory—believing in yourself more than in God. All along he is deaf and blind to his unconscious motives, which play out on stage for all to see. *The Tower of Babel*—a symbol of human pride—is by Pieter Brueghel (Kunsthistorisches Museum, Vienna). If somebody is truly arrogant in this chapter, it would seem, it must be Galileo, not Bellarmine. The origin of Bellarmine's portrait could not be ascertained. The two self-portraits are by William Utermohlen, after he was diagnosed with Alzheimer's dementia. Apart from historic inaccuracies, the chapter is flawed by an excessive insistence on the number 7: the 7 last words, the 7 deadly sins, 7 months, 7 tomes, 7 letters of the alphabet, and an inordinate number of sentences of 7 words.

25

Nightfall III: Dolor

*In which is said that, if the quality of consciousness
is a shape made of integrated information,
it can be turned into the only real and eternal Hell*

"This is a machine like no other," said the Master. He was standing in the dark door frame, a thin man with burning eyes—wearing a hat the like of which Galileo had never seen. After making a polite gesture, he led Galileo into a large room paneled in thick wood. At its center was a pedestal surmounted by a structure that looked like a catafalque, with the top covered by a purple drape. Above the drape a cantilevered brass arm suspended an elaborate contraption of interconnected parts, kept together, it seemed, by an extraordinary number of springs of all sizes. A sharp needle made of gold and glass was slowly moving up and down, tied to an oscillating hook, and then vibrating briefly, almost unnoticeably. Seven accessory needles were delicately pivoting around the central pin.

"I know the embroiderer never fails to impress," said the Master, throwing a casual glance at the machine, "but that is usually for the wrong reasons. It is, after all, just a machine. What matters are the instructions that it follows, and what matters even more is what one does with it, although of course the machine must function perfectly."

Galileo could not stop looking at the moving parts, and the Master took him aside to show him something he kept close to his chest—it . looked like a blueprint drawn on a copper plate. "This is how I set up the embroiderer," the Master said. "The most important aspect, though it's rarely recognized, is to obtain an accurate map of the meshwork, and from case to case the meshwork is always different. I need to know how the threads are stitched together, all the nodes and all the knots. This one meshwork is simpler than most, not surpris-

ing, perhaps, considering where it comes from. I wonder if it was worth my time—at any rate, the meshwork map is the most demanding part. But when the mesh-work map is done, I can begin to draw the sketch, and prepare the instruc-tions: which threads must be rewired, which new knots tied or strength-ened, which loosened or untied, and finally which nodes must be pressed, in what order, and at what speed. Then I can play the original instrument, so to speak, and make it sing as loud as it can bear, like trying the voice of a powerful organ, careful not to break it."

Galileo went back to the catafalque, to have a closer look at the needlework. There was a square opening in the purple drape, but it was difficult to see what lay underneath, as the moving needles formed a vibrating fence that blocked the view. What is the purpose of the machine? asked Galileo.

The Master stood still and clenched his lips together. Then, strangely emphasizing every single word, he said: *"What is the perfect pain? Can pain be made to last forever? Did pain exist, if it leaves no memory? And is there something worse than pain itself?"*

Galileo tried to hide his surprise. "I see you don't understand; how could you?" the Master mumbled. "You know the authorities well enough, don't you? All they want is a confession, and that I provide, I always do, what could be easier? Little do they know that it takes no effort to break a man, and no time." The Master waved his hand vaguely away. "To make someone confess is nothing, there is no art in it. The authorities," he insisted, "they only care about how many they can flaunt—confessions, conversions, recantations, repentations. This year the Cardinal of M. had two hundred thirty-one, that of

F. ninety-one, six hundred forty the one in D. But they don't care for how it's done. Once you have seen one, you have seen them all, they think—they are practical men. The truth is, nothing leaves any impression on these Philistines, save rank and money. Alas, my art could not subsist without their commissions. Even the Master must satisfy their vulgar needs—*every steed has lice, and every artist has patrons.*"

What is beneath the drape? asked Galileo.

"You are a scientist," smiled the Master as if relieved—"nothing that's human should be foreign to you." And he carefully unbuttoned the drape and lifted it from the catafalque. A man was lying on a butcher block, his ankles and wrists tied down with irons, his scalp exposed.

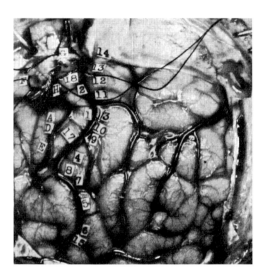

"You see," said the Master, "the man must be strong and well fed, this one has been at the embroiderer for more than seven days." The Master put the copper plate inside a slit at the base of the embroiderer and turned to Galileo: "These dogs will never understand. They want crude pain, common pain, *pain like gravel.* They fail to see there is no point to it, there is no point to their confessions, they mean less than the braying of an ass." The Master cast his eyes down: "They only know the crudest way to truth, so crude that truth is crushed by the uncouth surgery."

Not knowing what to say, Galileo was staring at the needles, which were probing inside an opening in the skull. A large mouthpiece pushed into the man's teeth, but the face betrayed no expression. The Master looked first at the embroiderer, then at the condemned, and then addressed Galileo: "You see how lightly my embroiderer spills the lightning flash of pain? I can deliver that with not a drop of blood—the merest spark between its pins and the sensitive tips of the nerves inside, and the brain flares into a flash of perception. Like you, *I seek the essence of perfection: the paroxysm of pain, the purest, the most powerful, penetrating, painful pain of all.*"

Galileo's mouth was dry, and the Master offered him some water. "It took me a long time to reach this stage," the Master said wearily. "For years I traveled, consorted with foreign medics, studied the ancient knowledge of the body, but none of it was of much use. And then one day I learned my most important lesson: *that pain inflicted to the body is a mere shadow of that which can be born out of its sources in the brain.* My helpers had been carried away with enthusiasm, and by mistake a needle pierced through the temple of one of the condemned. When I had gained his trust by feeding him some boneless fish, he told me this: there was no pain like it, none could compare. That day changed everything. Others, ordinary explorers, may be fulfilled by finding the spring of some great river. I wanted the spring of pain, the one that spouts inside Everyman's head.

"But it's a lonely art," the Master whispered to Galileo, as if sharing a secret. "Nobody to talk to, no disciples who share the knowledge and the passion. It may be hard for you to believe it, but the only solace comes from some of the condemned: by the time I am done with them, we have become quite intimate, more intimate than any friendship would afford, and the brightest of them open their eyes, so to speak, and see what I am after, they understand, and when they do, the search becomes excited, we tour each corner of what they can perceive, they even guide the embroiderer with their own eyes, or so I like to think. At last, alas, when they are numb and useless, *I must pour tired confessions out of the trite pain of the evening* (the torpid evening, when the senses lie tired and muzzled)."

Galileo remained silent, so the Master went on: "At one point I despaired, I felt that the search for perfection was bound not to succeed, doomed by the imperfection of the flesh. Whenever I came close to the purest pain, my subjects failed me, the residual tension in their limbs relaxed, their eyes became transfixed, *they hid from me the sizzling of the truth.* Did they achieve the highest pain, did I achieve it, and was it lasting? I could not tell. Does such pain inevitably wane, I wondered, dulled by numbness, fatigue, and sleep? *Nerve matter melts, flesh is too weak, the truth dissolves when it is burned by pain.*"

They both looked at the condemned: he had not moved at all and was holding fast to the mouthpiece. "Surely you understand me," the Master added. "Though the instant of pain may be the most intense, that does not guarantee perfection: the culmination of intensity is negated by the insubstantiality of duration. Thus in my latest work," the Master said, pointing at the condemned, "I have resolved to make pain reach true perfection and give it what it lacked: I have resolved to make pain unending. And now I have it—a pain that lasts forever at peak strength, unfazed by human failings." Saying that, the Master stood up straight and lifted his arms above the condemned: *"Perfect pain must last eternally, consciousness must rise to the point of sharpest suffering, that point must stretch to a line reaching infinity.*

"He cannot hear me," the Master said. "His mind is preoccupied with pain. But I see that you are curious," he said, peering into Galileo's eyes. "Of course I know what would be best: replace each nerve with one made of steel and glass, and keep the machine of consciousness running at full speed forever, spinning out pain, so to speak, as if it were its only word. The time for that will come, an artificial brain that can be tuned like a golden organ (oh, it will come, and somebody will tune it to the perfect pain), but I found a simpler way, bathing the nerves in a solution of arsenic and silver. *So I have made them invulnerable, the nerves connecting the high pulpit of pain to its entranced audience in the great church of the brain.* They will scream forever when the embroiderer touches their heart." In saying this, the Master accosted the condemned, took the mouthpiece and the wet rag from his foaming mouth, and halted the regular hopping, shuttling, and clinking of the embroiderer. He waited a few moments, and then whispered something in the ear of the condemned, delicately touching his shoulder. The condemned breathed heavily and spit something out. Then he thanked the Master in an exhausted voice, looked at him meekly like a grateful dog, and asked for some water. The Master helped him to drink, placed the rag back into his mouth, checked the instructions on the copper plate, and restarted the embroiderer.

"You may wonder why he is not even screaming," said the Master to Galileo. "Perhaps you wonder whether any of this is true. The man does not say he is in pain, after all. Indeed," he added, "medicine sometimes has its uses. You too should have a taste of this distilled spirit"—and he lifted a bottle to the nose—"which numbs the memory but not the pain. I tried it first upon myself: the pain was there, intense as ever, perhaps even enhanced, and lasted long enough that I could write down my sensation, but a few instants later I had forgotten it. Had I not written so, I would have never known that I had been in pain. Since then every condemned has received the spirit: I can turn on the embroiderer a thousand times," the Master said, "and when it stops, the man will smile and thank me. *Pain without memory is like the perfect crime.* It makes you wonder," he said with a thin smile, "the most acute of pains may well be the peak of reality, but does it really exist, if it leaves no record? The instant of pain could be the most unbearable, be followed by countless identical, fulminating instants (*like a migraine that made the head wish to be split from the body, the eyes roll on the floor to thank the ax*), but when it ceases, it has not left a trace, and nobody is guilty."

The Master went to check the embroiderer—for an instant the expression of the condemned seemed at odds with the template inside the machine. Satisfied with what he saw, he turned to Galileo. Not that he was content with pain, he said. True, pain had been his guiding post, his beacon, and the first spring to quench his thirst, but discoveries rarely come alone—*there are always other women after the first, and other sins after the original one*. So he had searched for the fountain of dread, of emptiness, and of envy, and found the chords to play to evoke them. Indeed, which were the true dimensions of human consciousness? For years he had probed and mapped the vast continents of perception, seeking its nodes, as if they were its cities, its mountains, and its rivers. In his chamber, alone with the condemned, he had fired up their intimate sensations. It was hard to believe how poor some people were—some could not distinguish between horror and hatred, or between rage and fear. To some even the colors were the same. But others had revealed to him the strange sensations that can't be triggered by the senses, and those that cannot be described in words.

The Master knew he must go further. It was not enough to map the continent of sensation: he must beget new peaks, peaks that had not yet existed, volcanoes rising out of the sea, a sixth sense, more exquisite that our modest complement, perhaps a seventh one must knit a meshwork of unprecedented design.

Galileo remembered Alturi, when he had said that by changing the shape of the quale, one could create new realms of sensation, though that was something he had not understood. The Master had no need for theories. He knew any map could be refined by learning, new shadows of sensations could arise—in a child growing to discern the novel taste of wine, a poet savoring all flavors that melt inside a word; but those were mere variations on a theme: now that he had mastered the meshwork over which the architecture of consciousness was built, why not mold it and warp it at his will, creating new pinnacles of suffering, labyrinths of darkest gloom, pits of inverted hope, ever-expanding explosions of lacerating loss? And what was the opposite of perfect pain? Did it truly exist?

"This is why I built the Great Embroiderer—it is my masterpiece," said the Master, pointing at the ceiling and watching Galileo closely. "It does not just poke in the nerves, but stitches them together, weaves them into intricate creations, *ties novel knots in the brain's enchanted loom.* If it is the only true theater in the world, why should I not transform the stage of consciousness?" The Great Embroiderer was coming down slowly from the ceiling like a sparkling chandelier, with calipers and scales and tiny wheels, and a forest of acuminated needles that were pulsating like the antennae of a nervous insect.

The Master moved away from the condemned, glanced at the embroiderer, then suddenly began to undress, exposing his body, but strangely he kept his hat. It all went very fast: he did not look at Galileo, but went to the back of the room and stepped onto a table of white marble that seemed an altar. He lay down carefully, adjusting the back of his head into the copper trough. He moved away disdainfully the mouthpiece covered with a rag and did not drink of the spirit that dissolved memory. At last the Master took his hat off, unbolted a golden plate from his skull, as if he had done this a hundred times, and threw the plate away onto the floor.

The embroiderer was descending toward the shiny surface of the Master's brain. "Now you will help me, Galileo," he said. "I need to find the apogee of consciousness, but this can't be self done: I cannot trust the mute expression of the condemned, I have to know myself. Ah, there it starts," he said, almost with a smile, when the embroiderer began to peck his brain. "In a few instants it will reach the center. Then, you will have to guide the throbbing movements of the embroiderer toward the point that is most sensitive, and I will tell you so. I will guide you to my apex, I will experience it in full: pain will explode, and its explosion this time will have no end. I am close," he said, "I never was so close"—his mouth was shivering—"it does not ebb and flow, it's steady, it's like the mounting tide that will drown all, but *I will swell high with that tide into the vortex, a vortex that gyres upward and not down.*"

He paused. A long silence, his lips stuck in disbelief, but then they mastered a whisper. "Alas, it is not pain," he said. "Ah, now it has changed. What is it? What is it that went wrong? That is it then, I know it now. When it is most intense, pain looses its identity, and when it comes to a point, the point is the same point where all else ends. Ah, no, it seemed that way, I know what it is now. Pure dread? No, not even that: pure loathing, that is where it ends—loathing supreme, of all things and of God. Loathing—the highest point of consciousness, loathing that overcomes the pain, and it will last forever. No," he said, "the embroiderer must have altered some connections, stitched the wrong threads inside the loom." Indeed, out of the incision on the skull, a tangled skein of nerves was rising, enmeshing the vibrating needles, mushrooming to envelop all the pins. "It is not pain, not fear, not dread or grief or loathing," he said. "What is it? Something new, unspeakably new. Why, I must have it now, is there no word, no there is not, but it is Hell; torture's a game, but this is Hell, the Hell that is most real, the Hell of consciousness itself." The Master's lips were closed, his eyes were staring like those of the condemned. Meanwhile, the embroiderer proceeded with its work.

NOTES

The inspiration for this truculent chapter is clearly Kafka's *In the Penal Colony*. Sure, the action takes place not in some tropical country but in Europe. (Babbage's Analytical Engine is at the Science Museum in London; Armando's *Kopf* hangs at the Dordrechts Museum.) Also,

Kafka's captain is merely an epigone of the old commander, whose golden age he constantly bewails, while the torture Master is clearly the commander himself. And his obsessive goal is not inscribing the definitive sentence of guilt on the condemned's skin, but exploring the remotest corners of consciousness. Naturally enough, the chapter lacks the inimitable sobriety of Kafka, often lapsing into baroque exaggerations. It has to be said, however, that there is some merit to a torturer who works directly at pain's source in the brain. Compared to that, Kafka's insistence on the gory details of how the skin is pierced and blood is spilt is rather superficial, just like the embroiderer is a more refined conception than Kafka's harrow. The torture artist is also more creative than his counterpart: he has actually figured out that the embroiderer can be used to stitch together a modified meshwork of connections in the brain, giving rise to new shapes in the space of qualia: new sensations, never previously experienced, still without a name, but predicted by the arcane considerations on the quality of consciousness in "The Palace of Light."

As far as one can understand, this is what the chapter is trying to say: (1) The most intense pain is evoked by stimulating the brain, not by torturing the body. Indeed, the chapter seems to indicate that pain, or for that matter any other sensation, can be experienced by the brain in isolation, without the collaboration of the body (unlike in Kafka). This suggestion is not as absurd as it seems. One can dream of pain without anything happening to the body. Also, some of the most intractable forms of chronic pain are triggered by malfunctions within the brain, not the body. A more difficult question is whether experience (of pain or anything else) can truly be completely "disembodied." In *The Feeling of What Happens* (Mariner, 2000), Antonio Damasio argues otherwise—at its heart consciousness is consciousness of the body and of its interactions with the environment. But both the Master and Galileo seem convinced that once an appropriate network of connections is stitched together—one that can give rise to the appropriate shape in qualia space—that is all it takes to generate consciousness. (2) If one were to substitute for neurons of flesh artificial, virtually immortal ones, as long as their interactions specify the same shape in qualia space, that very pain would be experienced forever, without fadings, faintings, slumberings, and distractions that inevitably enfeeble natural pain (replacing neurons with chips one by one is a classic thought experiment suggesting that consciousness is independent of its material substrate, though not of its informational one). Perhaps we should soon begin worrying about inadvertently creating machines that may undergo unpleasant experiences. (3) A surreal possibility is that acute eternal pain may occur without leaving a trace when memory is erased, as is done by some anesthetics. (What kind of world would be this, in which unspeakable sufferings go on forever inside each private consciousness, but nobody can report about it, not even the subjects themselves?) (4) Any other conscious sensation can be produced by stimulating the brain in the proper way (though we still lack an appropriate embroiderer), and even our maps of the connections among neurons are still inadequate; the exposed surface of the cerebral cortex is from Penfield and Boldrey, *Brain* (1937), who systematically stimulated different areas to see what sensations were elicited. In the last century, two psychologists, Titchener and Külpe, thought that consciousness could be decomposed into its atoms and, like the torture artist, set out to find its table of elements. The trouble is that Titchener found 44,353 elementary sensations, and Külpe fewer than 12,000. The larger trouble, of course, is that consciousness is not made of atoms, as discussed in "Seeing Dark," and the perception of pain, like that of dark, or any other, requires the large repertoire of the entire neural complex. (5) Not even acute eternal pain is the worst nightmare (though a rather dire prospect): worse horrors lie in store if one were capable of molding the fabric of neuronal connections, and thereby the shape of qualia, to give birth to new, abominable sensations. If it becomes possible, somebody will be willing: what was once thought can never be unthought. Of course the Master could have tried to create sensations even more heavenly than bliss, but then he is a torture artist, and this is "Nightfall III."

26

Twilight I: Consciousness Diminished

*In which is said that consciousness can be present
in the absence of language and reflection*

He had become an ox one winter night. The poet had fallen on his
verse—fearing his voice would never touch the ground.

Like dogs pursuing a wounded prey, they chased him through the
streets and stormed inside the house. He fled over the roofs, with rage
choking his throat: justice had fled with him, he thought, the long
night of the soul had dawned.

THEY HAD ACCUSED HIM: A CASE OF PERSONAL FEEBLENESS AND
THOUGHTFULNESS HAD BEEN ONGOING WITH HIM, AN EXCESS OF
SELF-CONSCIOUSNESS HAD DRAINED THE ENERGY OF THE BODY. THE
VANGUARD MUST ERADICATE REFLECTIVE ELEMENTS, THEY SAID, OR
THEY WOULD DRAG DOWN THE GENERAL TEMPO OF LABOR, AND SABO-
TAGE THE STRIDE OF THE MARCHING MASSES.

But he had thought: His life must surge like the morning sun, to reach a noble noon in Heaven; no, his eyes would never contemplate the evening, his star would not set quietly, but drown into the sea on wings of wax—he would soar to the cusp of his circumference, to tame the arguments of falcons, to peer into the pupil of the sphinx.

THEY HAD CONDEMNED HIM: TOADIES LIKE HIM BURDENED THE TAIL END OF THE WORKERS. WHAT HE CALLED THINKING WAS JUST A RANCID KIND OF PLOWING. HE PLOWED HIMSELF INSIDE HIS ROTTING BODY, SPURNING THE COLLECTIVE CONSCIOUSNESS OF PEASANTS. THINKING LIKE HIS MUST BE EXCISED FROM BRAINS——A CLEAN CUT OF THE SCYTHE INSIDE THE SKULL; AND HEADS LIKE HIS SHOULD BE EXCISED FROM THE NATION, SO THE GENERAL GOOD WOULD NOT BE SPOILED.

But he had thought: Why should it end before the morning time? No one had seen him fly, he had not seen the sea. Perhaps, to instigate a brand-new world, the discordant, tangled days must be linked, as with a flute. Brothers, let us glorify freedom's twilight—the great, darkening year, he had said. But to whom? To his own comrades? Who had laughed and wept and drunk with him, and then betrayed friendship and called him an enemy of the people? Or to the indignity of this wretched time?

THEY HAD DECLARED: THERE IS NO PLACE FOR PRIVATE SENTIMENTS IN A GLORIOUS NATION, THERE IS NO INTEREST FOR INDIVIDUAL FATE IN A REPUBLIC JUST AND EQUAL. CONSCIOUSNESS MUST BE ORGANIZED, AND CADRES ARE IN CHARGE OF THE ORGANIZING. INDIVIDUAL THINKING IS A FLY IN THE OINTMENT OF PROGRESS.

But he had thought: Can those conceiving the general and the abstract grasp a single man's soul without taking him prisoner? Those who concoct categories celebrate the twisted path of individual fate? No, they would march on, to victory, their minds carried in the stream of the identical present, identical for all, too fast to let them think of where, and whence, and why, waters too deep to let them taste their tears—deeds were too eager, and thought was lost in action, in this wretched land.

The poet had thought of this when his foot slipped and took him down, falling over the iron fence; the spike slashed through his brow from side to side. And thus he had remained, transfixed in midair.

The peasants had been wise, doctors had said. They took the fence out of the poet's brain, but trouble was, the brain would not let go of the fence. His brow had been gouged empty, but they had made it heal. He had never lost his sight, though he had lost his speech. They took him into custody and sent him far away. They said he was alive, but from that time forth he never spoke again. His sister in poetry stood for him, in front of a locked gate, she had stood for him for three hundred hours. But never did the doors unbolt for her.

Then, one warm spring day, they had released him out to graze. That was why they had unshackled him: dumb and docile he had come to her, and when she wept, he had licked her hand.

But it was not to soothe her. She learned it later, when his father died. When he saw him lie in state, he walked around his body showing no interest, as he used to do with things that were not new. Instead he wandered off, to mouth glass trinkets between the rude lips of an adult animal. And when his friends had been caught and exiled, he saw them led away in chains, but nothing touched him save his food. He led his dog in chains, or his dog led him. Oats in the cup, that was his only prize, and if there were bits of flesh he would grin and smack. Yet when she stroked his neck or chin, he would drone for hours on end. And he would fetch things for her—buckets and books and stones—he would not tire of it.

He could not speak because his frontal lobes were gone, the doctors said. He could not reason much, they said. His judgment was no wiser than a bird's. He pecked what was at hand and buried it within himself. His countenance was lost on him: no mirror would return his self to him.

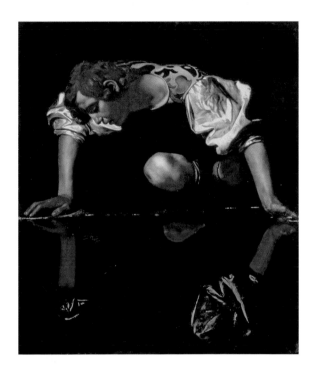

Galileo bowed when the woman rose to greet him, tightening the proud arc of her back. The river of her life had been diverted, she said, but not her inner course. Let them stifle her work, she would still sing his lot—her mute brother in poetry, his words would sound inside his people's mind. Though they had spayed him of his self, his verse they could not silence. And people would commit her mourning chant to memory, bear witness to his art and to the crimes of history. Her wretched land was a murderer of poets.

So she had endured, weighed down by doom, fearing each man and woman, known or unknown, fearing the news in each and every word, each new proclamation, dreading the number of each uneventful day, reading her verdict on the writhing lips of clerks. Gasping, the future stuck in her throat, she watched him breathe unencumbered, approaching everything and everyone tamely, sniffing and lapping at his slice of present. Her past was full of wounds and graves, still screeching in her skin; his was like a pasture, empty save for the sweeping wind, nothing to hold his sight.

In the stern of his head the brain was flickering in smoothly changing patterns. Like a placid film breezing over the surface of his cortex, scenes changed, events unrolled, life elapsed, and consciousness flowed. Shapes and colors, faces and places, sounds and noises, the shears and strains and flutters were still all waving there, just like before, but there was no catching them, no turning them inside the head, no thought of what they meant for him. Because at the fore of his head a chasm had opened, and at the rear the kaleidoscope of images had lost its mirror, all reflections broken—the back had lost its front.

Like a child who darted through the woods pursuing a hare, a child hurrying in the immediacy of the present, with no thought or reflection save for pursuit itself, slipping through the branches and jumping over roots and stones, too fast to ask or ponder, innocently without its self, just rushing through the fast-paced movie of perception, he was like that, just slower, when the night found him, foraging tardily over the fields.

NOTES

For some reason, "Twilight I" begins with references to Russian poets and writers. "To instigate a brand-new world, the discordant, tangled days must be linked, as with a flute" and "Brothers, let us glorify freedom's twilight—the great, darkening year" are verses of Osip Mandelstam, who died on his way to Stalin's labor camps. (There is no record of exactly how he died, but the story of him receiving a wound to the front of the brain while he was trying to escape must be apocryphal.) Other references are to Mandelstam's friend the poet Anna Akhmatova (*Requiem*), whose first husband was executed under dubious accusations, and whose son was interned in a gulag. The paragraphs in small caps must be inspired by the work of Platonov, especially *The Foundation Pit*. The iron curtain (modified) is from Paul Artus's web travelogue. The central theme of the chapter is that, after extensive frontal lesions, self-consciousness and reflective thought are reduced, but consciousness remains. Frontal lobotomies (the scythe, the sickle, the iron curtain that pierces through the frontal lobes of the brain) were all too frequent in the first half of the twentieth century although, it should be said, not in the Soviet Union. (In that case, it may have been the nation as a whole that was lobotomized to suppress criticism.) A famous case of a transfixed head is that of Phineas Gage, a railroad worker who in 1848 inadvertently blasted off a tamping rod that rocketed through the front of his head. The well-mannered Gage survived, but his personality and behavior changed much for the worse. Hanna and Antonio Damasio showed that the rod most likely damaged the ventromedial regions of the frontal lobes (Damasio et al., *Science*, 1994). A similar case was a Spaniard impaled by a spike of an iron gate when escaping out of a window during the civil war in 1937 (Mataro et al., *Archives of Neurology*, 2001). A pure example of global frontal degeneration is a German lady described by Markowitsch and Kessler (*Experimental Brain Research*, 2000). There is now neuroimaging evidence that one can be conscious without being self-conscious, in which case the back of the brain is active but the frontal lobes are quiet (Goldberg et al., *Neuron*, 2006). The first painting is Pieter Brueghel's *Landscape with the Fall of Icarus* at the Musées Royaux des Beaux-Arts de Belgique (Brussels). The second is Correggio's *Zeus and Io* at the Kunsthistorisches Museum, Vienna. Zeus wanted to seduce the nymph Io (in Italian "Io" means "I") and transformed her into a heifer to avoid Hera's ire. The third painting is a detail from *The Earthly Paradise* by Jan Brueghel the Elder, at the Louvre; the fourth is Caravaggio's *Narcissus* (somehow the reflection of Narcissus's face has been erased), kept at the Galleria Nazionale d'Arte Antica, Palazzo Barberini, Rome.

27

TWILIGHT II: CONSCIOUSNESS EVOLVING

In which is said that animals are conscious, too

Closer to us or far away from thought, creatures of habit, whose silence gives consent, do they share of darkness as of light?

Unknowing, the ass was hobbling up the shaft, the sacks of coal hung down its sides, the knees shaking under the weight. Behind was the dark hole of the tunnel; in front the lantern of the mule boy swung with each step.

Nobody knew why an ass had ended up down there. Mules were sturdier and could take cudgels all day. But the Hide, so they called it, had a skin harder than leather, as if they had tanned it alive: it did not budge when they hit it with the shovel, though at times, when the blade carved into its flesh, the neck would shrink into the shoulders.

Nobody remembered when it had arrived at the mine. An old jennet that might have been its mother had been there, too, they said. The Hide may not have seen the light of day, not even when it came out of the belly. Its eye was spent, and maybe it could not really see. It would not need to, anyhow, since it had always gone the same dark trail. It had just gotten longer over the years, as they dug deeper and deeper into the earth, or perhaps its gait had turned lazy. It sure looked wasted, and there was no way the Hide could last much longer. When the jennet had died, the Hide had stopped awhile at the pit where they threw the carcasses, and for hours they could not get it going again, no matter how hard they hit. Maybe the Hide thought that, now that the jennet was carrion, it would feel pain no more. Now that wild teeth and beaks were ripping it, the flesh would feel nothing no more. Maybe the Hide hoped that soon it would feel nothing either.

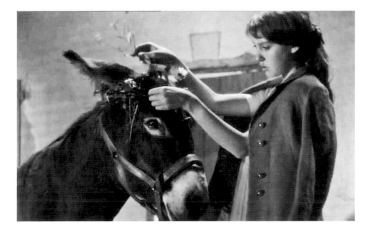

And yet the mule boy felt for the ass. It carried its pack every day and had never been sick, it had never even brayed: the boy had wondered whether the ass was mute. Sometimes he put his hand over the harness to see that it was tight, and then he stroked it as if it was his brother. The ass would turn its head and stare at him and rub its muzzle at his side. But now he only stroked it when they could not see him. They had laughed and told him not to spoil the ass. The ass had it no harder than them Christians, they had said. And it did not even have to worry about food on the table.

When the day had come, the boy had tried to hide it from the others. Up the steep part of the trail, the ass all of a sudden halted; then its front legs gave in, arching its back. The sacks slid down over the neck, and the boy worried the coal might slip all out. He had whispered something in its ear, it had to stand up soon, or they would have noticed. But the ears kept down just like the knees, and the knees were bleeding. When he tried to pull it by the neck, they saw him and called him off. So with the coal dust wet on his face, the boy had asked them to let him take care of the ass outside the mine. He promised he would get it back to work, and it would cost them nothing.

It was the end of the day, and nobody wanted to go down to the pit and take care of the ass. So they tied up them both, the boy and the Hide, and hauled them up the shaft.

When they got out, the sun was setting over the valley, and the shadows were chasing the light over the meadows. Without the pack saddle, the donkey was walking slowly behind the boy. His eyes were blinking like he was in a frenzy and could not keep them open. The head was shaking side to side, and to the boy it seemed as if the color of the skin had changed. Before long they were out of sight, behind the line of trees into the pasture.

It was darker now, and the sky too was a different color, and the wind was up. Then the donkey gave a brisk swerve, which took the boy by surprise and made him drop the harness. There was no way to hold the donkey back as he careered over the open meadow. When he was far enough, he kicked the air like a drunk, and made new noises the boy had never heard. It was as if he was not a donkey anymore but thought he was a bird—the way he kicked and jumped, he looked as light as one. It was as if he was greeting the sky, and then the grass. It was as if he told his limbs they could still run and spring unburdened. It was as if he wished to breathe the evening air to clean his lungs of dust. It was as if he was crazy with joy.

"So," said the old bearded man to Galileo after a while, "you have seen a human ox and a donkey human. Whose pain and joy was more intense? Or was there no experience behind the mask? Were they just rusty machines feigning a soul?"

Galileo was sure of what he thought. There was no reason to deny consciousness to the human ox, any more than to himself. He might not contemplate his own reflection in life's mirror, might not ponder the burden of his actions, or might forsake deceit's sharp subtleties, yet certainly the river of experience still flowed inside his brain. It was just the ambiguous northernmost meanders, the frontal lobes where the reflecting self dwelled, that were cut off from the main course. All humans were such oxen, then, when they awoke from a deep slumber, or wanting for lack of sleep, or even in some dreams, but then experience was not extinguished, only straightened and simplified. So when the brain's back was not reined in by the stern harness of the overbearing front, not by accidents of fate but by the vicissitudes of nature, as in the donkey and countless other animals, what was so different from the human ox? As much as they were similar in their behavior and their brain, why should they not be similar in their experience? Perhaps not quite the same experience: a man's stiff gait was not the leopard's leap, the darting of a dolphin from the sea, the bat's sharp-cornered flight, the octopus's smooth jointless sliding, but it was experience nonetheless. Was the donkey's freedom dance no dance, and its joy less real than ours?

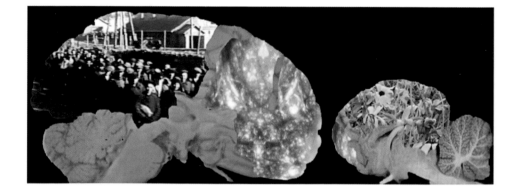

"Quite so," said the old man. "Their faces betray their feelings, just as ours do. More so: our feelings betray their feelings, and our envy is their mutual struggle introspected, our jealousy their open fight for mates, our love of art a chimpanzee watching the peace of sunset."

"But tell me," the old man continued. "Should the crushing of an ant fill us with tears? Would its pain be as intense as the Master's pain in his embroidered Hell? And how would we ever know?"

So after Galileo had thought about it, he remembered the Frenchman he had awakened with Frick, and answered thus: When after two sleepless nights your mind surrenders and falls into the deepest of deep sleeps, and yet ungraciously I force you to rise and ask you whether any image or thought was occupying your consciousness, what would you say?

"Too many times I was awakened from the deepest sleep," replied the old man. "Nothing was on my mind. I might as well have been like dead. Dreamless sleep is nothing but the abject annihilation of consciousness."

Perhaps not so, said Galileo. Perhaps a whiff of consciousness still breathes inside your sleeping brain but is so feeble that with discretion it makes itself unnoticed. Perhaps inside your brain asleep the repertoire is so reduced that it's no richer than in a waking ant, or not by much. Your sleeping Φ would be much less than when your brain is fast awake, but still not nil.

"Just as the temperature of the coldest air is not true zero, not zero absolute," said the old man. "And yet it is so cold that all life freezes."

NOTES

The mad joy of mine mules brought to the surface after years has been described in Stephen Crane's "In the Depths of a Coal Mine" (*McClure's Magazine,* 1894). The plight of boys and donkeys in mines was memorably told in *Rosso Malpelo* by Giovanni Verga (1878). The first painting of the donkey is by Johann Georg Grimm (Bühl am Alpsee, Bavaria, image kindly provided by the Grimm Vereinigung, Immenstadt). The black-and-white photogram is from *Au Hasard Balthazar,* the story of a donkey filmed by Robert Bresson. The second painting of the donkey is by Tino Vaglieri (Museo Civico Bodini, Gemonio, Varese). The donkey and other animals in the sky are a *Capriccio* by Goya (Musée des Beaux-Arts, Agen, France). The last image shows the relative size of the human and the equine brain. Note that the prefrontal cortex (the broken mirror) is much larger in humans. In *The Feeling of What Happens,* Antonio Damasio has argued that a primitive sense of self, closely tied to the body, is the most ancient foundation on which animal and human consciousness rests. This protoself would certainly be present in both the donkey and the human ox of the last two chapters, whereas the autobiographical, reflective self that animals may lack, and men may lose, does not appear to be a necessary condition for being aware.

28

Twilight III: Consciousness Developing

*In which is said that consciousness must be present,
to some degree, even before birth*

There was no one in Heaven—no one to judge between life lost, and life won and unwanted. Crumpled over himself, his head tucked inside the chest, the Pope's physician was resting next to the girl's splayed limbs. The girl was covered with a purple cloak, her breaths shivering faintly out of her middle body. On the floor the bowl was full and the forceps soiled.

There is no need to hide, said Galileo, any more than I would have to, doctor, as there is much we share. Besides, what is the crime this time, hypocrisy or murder? It seems to me, though mine is no trained eye, the girl is going to live, and she may yet gain from the present loss—she will soon thank the eminence of her surgeon.

Seeing that he had been diagnosed, the Pope's physician raised his head to speak. "Oh no, my friend, of course not. I need to explain my circumstances, though excuse them I cannot. It was a deed I did not want my hand lent to, but it was called for by a higher cause. If every seed of human weakness were allowed to ripen, the stench of its foul rotting fruit would wilt the brightest flowers of the Church. What lessons would the young and pious derive? The grand edifice of faith might shake for the sake of an unformed clot. So it must be done, done by an artful hand, a hand affiliated by blood and acquainted with discretion."

Then why do I see your spirit burdened, doctor? asked Galileo. Was this done after the fortieth day?

"I see you are missing irony's sharp blade," said Rome's Protomedicus. "For I was the very one who espoused the infusion of the soul at the precise instant of conception, till it became the doctrine of the Church. I was the one defending the authority of science and the straight line of logic, who proved that body and brain are formed in a seamless process; I was the one to whom electing the fortieth day, or twice as much for the female sex, was a date arbitrary and incongruous in the extreme; the one to whom forcing an artificial birthday onto the magnificence of the rational soul was an irrational non sequitur. But now, my friend, if it is life that proves non sequitur—why worry about consistency in the doctrine?"

Then you do not trust the poet, when he gave words to thoughts of great philosophers? asked Galileo:

Open thy breast unto the truth that's coming,
And know that, just as soon as in the fœtus
The articulation of the brain is perfect,
  The primal Motor turns to it well pleased
At so great art of nature, and inspires
A spirit new with virtue all replete,
  Which what it finds there active doth attract
Into its substance, and becomes one soul,
Which lives, and feels, and on itself revolves.

This is what I read into his verses, Galileo went on: certainly there must be a time at which a new life begins, and that may well be at conception. But plants live, too, and yet it is no sin to take their life away. There must be a time at which sensation starts, but even that makes us into no more than animals, which are not spared, whether born or unborn. And finally there comes a time when reason glows, and knows it sees the world, and discovers the self. That is the time when the full soul is born. Whether that happens on the fortieth day, as they once said, or when the infant opens his mouth to speak, or at the eighteenth month, when it learns of his reflection in the mirror, I don't pretend to know.

"That will not do, my friend," said the Pope's doctor. "Like a philosopher contemplating the paradox of time, each moment divided into a chain of instants that is infinitely long, I am perplexed. Perplexed in front of a mere clot, which grows imperceptibly into the majesty of a human soul.

"I know full well within myself that the clot I killed was just a clot, and had less soul in it than the wart I removed from the Pope's brow. Just as I know within myself that by this one unreal murder, I may have saved the fate of one real life. And yet within that clot, the full trajectory of a man or woman was waiting to be born and grow, and to become one day perhaps as much as a new pope. Now that its course has dried, like a riverbed in which water will flow no more—begone without being—it will beget no life, no feeling, and no memory, it will not give and not receive, it will not even have a name."

The trouble with unborn life unfolding its potential, said Galileo, the trouble with all slippery slopes, is that slopes slip forward and backward, too: if there is no discontinuity after the pivot of conception, there is none before it either. Then the omission of emission is a mortal sin—tell the Congregation for the Doctrine, he added with a smile. And think how the daily genocide of abstinence makes murderers of us all.

"No, my friend," replied the doctor. "Before conception there is no novel being *in fieri*. Afterward there is—a being with the clear marks of humanity and the distinguished traits of a unique individual. I saw it with my own eyes: I saw the trace of a man's effigy at the embryo's earliest stages, ready to unfold into a full-formed person."

I shall not fight you over spent semen, said Galileo. A man named Frick told me that one may soon grow a newborn twin out of each

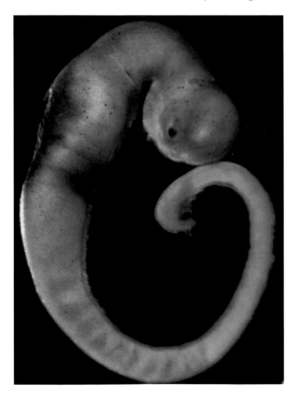

hair, so every shave is a potential holocaust of millions of your closest brothers—we'll leave this to the doctrine of the future. But the embryo is no full-formed person— it is a folded bundle of instructions. Though the architect may hold the well-etched blueprint of the great cathedral tight under his arm, there is no church until the church is built, a church that can be entered by the faithful. For bear in mind how we come into being: the seat of reason starts as a mere tube, which bends and bulges, then swells and thickens, and it grows eyes as if they were two fingers, pushed out of the palm of a reversed glove, then splits into layers, and lets its cells multiply and grow, wander, and spread their limbs both near and far. Then it begins to rule over its actions, at first clumsily and wanting of certain purpose, then of a single mind it starts to buzz to life, a tenuous bruise and then the loud cry of billions of nerve cells, at times awake and at times asleep, and then those cells turn into well-skilled experts, refine the circle of their handshakes, and cover their long arms with a tight sheet of insulating fat, to share fast shocks across their fingertips.

"I did not know, and I am humbled, that the academy of Prince Cesi had progressed that far," said the doctor. "But then what does Galileo himself think?"

There is less soul in a slumbering human embryo than in a poor old donkey when he tastes his freedom—this much I know, said Galileo. Early on, an embryo's consciousness—the value of its Φ—may be less than a fly's. The shapes of its qualia will be less formed than its unformed body, and less human than that: featureless, undistinguished, undifferentiated lumps that do not bear the shape of sight and sound and smell. So it will have little or no pain, hardly any feeling, and certainly no self. When should we grant it a proper soul? Does it just need to have lived, even for an instant? Or does it need to cry, or walk, to speak, or think, or ask, to recollect a memory of his own, or question why it was begotten one day, and wonder whether that was a good thing? It is fairer to reason that it merely needs to feel—to have even the faintest of sensations.

I do not know when the light of consciousness enters the growing cathedral of the brain. Does it do so by stealth, unnoticed at first, a candle lit in a dark corner of the building site, long before the baptism of birth? And does it then slowly light one chamber of the brain after another, until it comes to brighten its vast hall? I do not know. But the light piercing the world does not turn on with life—it flares up with consciousness.

"Be it as it may," said the doctor, shaking his head, "but though it may be feeling, and not just life that matters, how much feeling justifies a soul? The merest amount? Then there may be some feeling at conception, and how would you know how much there is of it? All we can judge is what it does, and that will not tell us what it feels. What if it were always dreaming?"

You know it yourself, dear doctor, Galileo replied: a being's repertoire of actions may well reveal its repertoire of feelings—diverse deeds or words, a rich internal universe; the monotony of stolid vegetation, a bark that's empty inside. Not an assessment we can safely trust, but it will have to do until we come to measure that repertoire inside its brain. But if you ask me whether a whiff of consciousness is enough to qualify for murder, I say tonight at dinner you'll eat a soul much larger than you just killed at the operating table.

"Maybe, and yet a whiff of consciousness in a human being can't be less precious than the donkey's joy. It is the only light that's lighted by a divine spark. Thus we owe respect to human life however, wherever, and at whatever stage," said the doctor.

Whether or not our spark is more divine, said Galileo, respect does not arise in things themselves, but in the thinking burden of our responsibility. Dear doctor, the mother of my children never was my wife, and prudence might have counseled that our fruits be snipped in their unknowing buds. But that was a less important choice than this: whether, having decreed their birth, we would commit to make those lives as full as fate allowed; that once committed, we would accept them for better or for worse, as we once did for each other. It is because we are human who understand, pass judgment, and decide, that we owe respect, and owe it to all that reason might find worthy.

"No easy prescription, Galileo," said the doctor. "The extent to which reason finds things worthy of respect changes with time, tradition, and knowledge of the world. And then it means that respect is not due absolutely."

Judging is never easy, said Galileo. But there is some wisdom in this view of things. Just as we owe respect to an old mindless man, or to a simpler animal than us, so we owe respect to a budding embryo. But we owe much more to him who the embryo will one day become: we should bring him to life and consciousness, only if we can hope with reason that he will one day agree to our choice. If forcing him onto the stage of birth is going to cause great pain, to him or those whose consciousness already shines, our innocence would be irresponsibility.

"You may be right, Galileo," said the physician, resting his head again within his shoulders. "But then what should I do, or have done? And what should guide our conduct? A rule of law, a principle of morals, or the truth of science?"

NOTES

The painting is Schiele's *Pregnant Woman and Death* (*Schwangere und Tod*) at the National Gallery, Prague. Paolo Zacchia (1584–1659) was the personal physician of two popes, legal adviser to the Rota Romana, and health supervisor in the Papal States. His most important work, *Quaestiones medico-legales,* gave forensic medicine its name and scope. On the other hand, there is no evidence that Zacchia performed any abortions. We already encountered the Pope's doctor in the chapter on the cerebellum, where he had diagnosed the affliction of the painter Poussin. The verses are from Longfellow's translation of Dante's *Purgatorio,* Canto XXV. The image of a neural tube (early mouse embryo) is from K.-K. Cheung et al., BMC *Developmental Biology* (2008). Prince Cesi was the founder of the Accademia dei Lincei, the first scientific academy, of which Galileo was a member. *The Carcass of an Ox,* by Rembrandt, is at the Kelvingrove Art Gallery and Museum, Glasgow, Scotland. *Saturn Devouring One of His Children,* by Goya, is at the Prado, Madrid.

29

Daylight I: Consciousness Exploring

In which is said that by investigating nature,
new qualia are discovered

"Why would one leave one's room to find Heaven and Hell?" It was a male voice that spoke, carried through a pipe ending in a horn of brass, hanging from high above the apartment door. "Exquisite fondness I have for those who catch in word or paint all that is seen and heard and felt, those who can pin on a virgin page every specimen of time past and present; photographers of experience, who miss no slide of life, plunge into the subterranean oceans of the self, capture, with consummate perfection, how purity is composed in childhood's Heaven, dissolved in the damnation of acquired wisdom, distilled in the purgatory of memory.

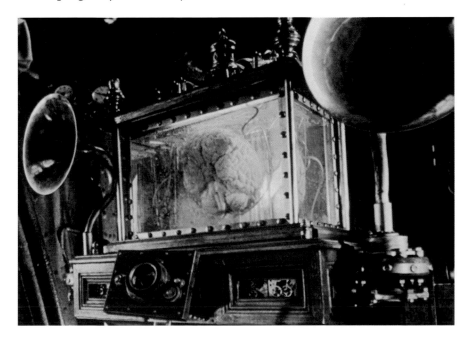

"Acquaintances told me of a painter who could do, with brush and oil, what I set out to do with pen and ink, and stream a swarming torrent of images from a room without a window. So from the training ground of my bed I have disciplined myself: each time the clock strikes a new minute, I compose a novel sentence. Sit on the mat and read it—Céleste will pass it out under the door. I'll give you a new experience every instant of your life: and with each sentence conjure an image in your mind, outwrite the pace of your tired eyes, outlive your life. For only a life relived in memory, replayed by imagination, unfurled against the sharpest light of thought, is a life truly lived." The plate next to the apartment door read: M.P.

Galileo stood in the landing, wondering to whom the voice was talking, but had no time to ask: a man swung open the door of the opposite apartment, ready to knock down the horn that spoke, or storm the neighbor's entrance. But seeing Galileo, he came to a stop and introduced himself. "Sir Ernest Henry," he said, "and all this suffocating talk blaring out of that darned horn is sheer and utter nonsense. Whatever the repertoire of consciousness might be, however wide and marvelous, he is just an abject beggar scraping its barrel—starving himself to stale words, spoiling the silence. A great writer, he says! Oh yes, the one who will expound forever a louse's unformed smile, half-unintended double messages in its bite, or circumnavigate each hole in the embroidery of his prison's curtains, and chronicle his introspective journeys till he and his musty self are out of breath in dusty spasms. Let me break through his door, let polar winds awaken his poor mind and freeze his lucubrations into frosty shapes that crumble with a sneeze."

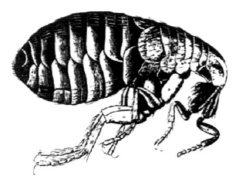

"Ah, my opposite has awakened," said the brass horn above the door. "I see you learned to speak the way I write, Sir Ernest. I told you many times, the voyage of discovery is not in seeking novel views but in having new eyes. The frozen landscapes of the Antarctic may well be icing on the earth's cake, but beware: the flavor of the cake is baked inside this room."

"How dare you speak to me of discovery! The quest for the new continent flexed our lives like birches in the storm, but we brought back new sights nobody knew to exist. What can you claim instead, bed-bound scavenger of life's gossip and trash? A new name for boredom? A lifelong collection of wasted hours?" And the explorer turned to go back to his room.

"Do not go away, Sir Ernest," said the voice. "If I won't pin you to my page, you and your ship will soon float by and vanish. The river of experience flows too fast, a vortex disappearing into the darkness of the earth, unless an intellect stems its current, and halts it, to record its passing. The steam of thought spirals upward, dissolves into the air, unless it can be caught—a net for butterflies is how you catch it—and crystallized, so it can be preserved. A gesture is a passing thing but needs a volume to be grasped in full—a volume that weighs like a life's work. You search new continents of ice, Sir Ernest, perhaps to preserve memories, but yours is a hopeless quest: a lady whose dress you cannot recollect, the flavor of her tea has evaporated, her wink has lost its special dart, is gone to dust, and cannot be revived. Even an age can vanish, when its smell is blown away by the icy winds of time."

"This is what literature does to you," mumbled Sir Ernest, look-ing at Galileo. "The writer's fate is sealed inside the pages of his book, enchained by sounds and syllables and sentences—a prisoner of words. But there is more to one experience than can be said in a million words. Had I to spend an hour, even a single hour, exchang-ing pleasantries with a bubbly glass at hand, a monocle on my eye, I would turn into a mummy plagued by an unbearable itch. The only gesture left would be to scratch it all away."

"I detect some haste, which is the enemy of depth," said the horn. "My dear Sir Ernest, let me scratch your itch. In any random photograph of experience, plucked from the course of life at any time, the entirety of its world must be contained—darkness is pregnant with a trillion scenes; and when kaleidoscopes unfold, one of their leaves is black. All novel lands, all undiscovered mountains, all views of sky and earth, far larger or far smaller than we are, already lie contained within the private repertoire of consciousness: Sir Ernest, all photographs of Antarctica were laid in consciousness's anthology long before you took them, and long before you saw them—nothing can be seen if it was not already in store—memory is imagination peering into the past, and imagination, memory looking out into the future. The world is like a curious child, turning the pages of our album and pointing to this or that: our pleasure is in discovering what we have, and sadness that we'll never know it all. But then your icy landscapes are not more precious than my dissections of a lady's smile—in fact, I'd rather snatch a kiss of her warm lips than stride above the cold lands of the penguins."

Perhaps the voice of brass is right, intervened Galileo before the explorer could—the repertoire of consciousness can be tasted as much by dissecting even a single impression as by sailing a ship to unseen images of novel lands. And yet no taste compares with the inrush of the new: day after day, for years on end, I traveled my eyes over the same sky, seeing each time a slightly changed angle of the same stars. I could have spent in elegant variation all my remaining days, or like a peasant silently welcome the turning circle of the seasons, contently telling the spring sky, the sprouts budding in the fields, and the quiet gait of the steer: I recognize you. But then I pointed the telescope toward the sky and a new compass rose, and saw untold unnamed things, and climbed high in consciousness's repository, higher than ever before.

"So it was for me," said the old bearded man walking up the stairs, "when I was bold and green, collecting every day a new exemplar of life in distant lands: if all of nature's images are just a chapter in the catalog of possible experiences, how varied a treasury does its index contain? But though the semblance of every species, the portrait of every individual, every single act and every sequel of the act, may be contained within consciousness's book, and vastly more besides those merely existing, countless pictures that could be imagined but never were or are or will be, yet when nature's true cards are finally dealt to us, and we have made acquaintance with the real, then a new expectant valley offers itself to the pioneer who, piercing though the verdant canopy on a lofty peak, can contemplate the expanse of his progeny."

After he had rested, the old man went on: "But don't be harsh with the brass horn, Sir Ernest: for exploration can be a sedentary affair. You can discover as much or more sitting at nature's desk, or at the table of the elements: architectures of earth and stone, divine geometries of fruits, singular landscapes of bones and feathers, the convoluted fabric of all life, untreaded countries in exotic brains, seas stocked with squirting sperm, the mighty engine of the cell, strands of ancestral memories, the jungle of reality that populates the repertoire of consciousness. Nature is never more complete than in its smallest parts."

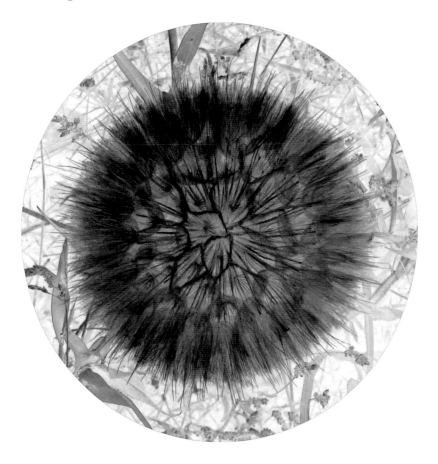

"How crowded and ebullient is my ward today," said another man, descending from the stairs in elegant morning dress. "Gentlemen, have you ever wondered what makes us males doggedly play the explorer's game? Why, we are all after the same thing!" He looked at the explorer, still standing on the doorstep. "Sir Ernest, what are you fleeing from? The shielding tent of your mother's skirt? It's dark in there, and like a migrant bird that rather has an ocean than a nest, who thinks responsibility is a prison and feels its wings are free only in the open sky, who travels always, to think he's never trapped, who wants to escape from the quotidian, because he really wants to escape from home, you crave an unclaimed portion of the air. And yet an albatross is just an outward version of a mole."

"Ah," rang the horn, "the incomparable Dr. F. has arrived. It's time for him to do his rounds, for us to bare our souls."

"Indeed," said the doctor. "But how does M.P. feel this morning? For he, our lettered mole, he too has learned to escape, if only inside," the doctor said, facing the others. "I told him so: too fearful to escape his mother's skirt, he burrowed his trail inward, searching for a safe recess, he crawled inside a corked corner to listen only to the calm beat of her heart, as he once did when he had no memory. Behold these two, M.P. and Sir Ernest," he said, turning to Galileo. "One turned the scope outside, and one inside, but both did so to look for a way out. Because both fear what's real. For what is real? The unnamed mountain that appears above the horizon, when your step touches the granite of its roots, or the fleeting expression of a flashing smile, which says what is unspoken, and perhaps unfelt? That which actually occurs, is conscious, shining in the foreground, clear and with a clear cause, which is explicit and can be understood through reason? Or that which might have happened, unconscious, lurking in the background, ambiguous and with more cause than one, implicit and perhaps irrational?"

"You see what kind of doctor we have earned?" exclaimed Sir Ernest. "Not the kind I'd gladly take along aboard my ship. But ask him, and you'll find that in his mind he is an explorer, too, maybe a conquistador."

"Sir Ernest is proud of his achievements, and rightly so," replied the doctor with a smile. "He sees us as usurpers, pretending to his candid bride, and cannot keep his cool—he has, so to speak, an Anti-arctic personality." Then he addressed Galileo and the bearded man with a deferential bow: "I, on my part, am glad to share the company of scientists, of those who search for what is true and real and deep. Because," he added, "if it did not sound conceited, perhaps one might conclude your humble doctor is quite a bold explorer: I too discovered new continents, vaster and richer than those Columbus found (Sir Ernest's barren bride cannot be in this league), what's more, continents that lie submerged like Atlantis.

"As for M.P.," continued the doctor, "he likes to flaunt his exquisite perceptiveness, but all he does is skim over the surface, describing every wavelet that curls the fashions of society. Yet he is oblivious, or should I rather say as blind as Oedipus, to the buried engine that holds all explanations—for every tempest, wave, or ripple that agitates the vast sea of the mind. But I uncovered to the peering eyes of consciousness the murky pulsions that stir the unconscious self, shed light where previously was night, disclosed sundry new twists to every image that the mind perceives, resolved uncertainty in the thick weave of memory's hidden strands. I sailed my ship in far more dangerous seas than Sir Ernest ever did: I brought geography inside psychology."

At which Sir Ernest could not hold back his laughter. "Excellent, doctor, truly an extraordinary performance—could make a convert of a stone. Not just the greatest of explorers, a sublime scientist to boot! Now I concur with you: science, too, of course, is preoccupied with lifting skirts, notably mother nature's. So this fixation of yours, on skirts and what lies under, may be a good thing. But when it comes to such a lofty matter, how sure are you that it is science and not imagination?"

"Science is imagination tempered by the real," said Dr. F., as if he had been called to judgment.

"No offense, dear doctor," replied Sir Ernest. "Is reality overdetermined, or is it overinterpreted? But never mind reality. Have you ever thought how much you are like M.P.? You both write well and both are haunted by feminine things. But I must say, as much as I resent my neighbor's blaring, it is poor M.P. who makes the better scientist: he may be a louse with asthma, he may be obsessed with trivia and social rank, but he at least sticks to the facts. Your story, doctor, I am afraid to say, belongs to the next chapter. Perhaps it's you who writes the better fiction."

"Well now, proud navigator," said Dr. F., "I feel your righteous envy: you spent your better days out in the garden, fearing the enclosed space of mother's house, searching for hidden treasures, but never found what you were looking for. And our M.P., as noble as his aim may be, revealed at most the hidden games of the boudoir, untangled the contorted thoughts that dwell inside the living room of society. Yet neither you nor he descended to the basement, to shed some light on the foundations of what is real. The kingdom of the depth is a large storage room, contains all toys and all ambitions, distilled by reason's cunning into the future path of men; clothes we once dismissed but hold truthful impressions of who we are, closer to the skin and drenched of our fragrance; the food on which the repertoire has feasted since its birth, and left a lingering in our breath no amount of drinking can erase. And yet before I shed light there, the largest chapter of the repertoire had not been read, though it lie written by the story of life."

"Well said, doctor," resounded the horn again. "All is well, we are all explorers: a mouse that never left its corner and never dared to venture in the world, what would you think of it? The repertoire is an endless book, a novel with a trillion pages, and would you want to read just the first ones, and maybe the last, to lie that you have read it all, and know what life is like?"

Then from a higher story a woman's song was heard, wafting like a breath of air inside their minds, before it vanished into silence:

". . . internal difference," she said, "where all the meanings are . . ."

Whose was this voice? wondered Galileo. Internal differences, where all the meanings are—differences internal to each experience, that made it what it was and different from a trillion others—the corners of the shape of qualia. He remembered the palace of light—the shining constellations specified by irreducible mechanisms—the lenses of the master grinder. No distinction within consciousness without an underlying mechanism, thought Galileo—without a difference that made a difference. Internal difference was internal meaning.

But who was she to speak about the space of qualia? He quickly climbed the stairs: a pale-faced lady was all he could make out through the door that closed—a special slice of Heaven glanced from a peeping hole.

NOTES

Daylight at last, but hardly any action in this chapter, just five white males meeting in the stairwell of some apartment building, probably in Paris: Marcel Proust, Ernest Shackleton, Sigmund Freud, and of course Galileo and Darwin. They all insist they are explorers, they all boast of some great discovery, and all do so in a pompous and long-winded manner. Despite the chest-beating about the relative importance of their contributions, it soon becomes apparent that the roads of discovery are many—discovery of new lands, concepts, and laws—and all can make us richer. At the end, Emily Dickinson (in the daguerreotype) saves the day with one of those pronouncements that show how poets (or women) have deeper intuition of what is real than scientists (or men) ever might: internal difference, where all the meanings are. As she reminds the explorers, discoveries about the world ultimately matter if they make a difference within one's consciousness, if they enlarge the repertoire of qualia and refine the concepts and distinctions within each quale. Said otherwise, exploring the shapes of nature ultimately means discovering the shapes of consciousness. And when she describes those shapes, she knows what she is talking about as well as any of them.

The brain in a vat is from the film *The City of Lost Children* (*La cité des enfants perdus*), directed by Caro and Jeunet (1995). Several philosophers have considered the brain-in-a-vat thought experiment, for example Hilary Putnam in *Reason, Truth, and History* (1981) and, without the vat, Descartes in the *Meditations of First Philosophy* (1641). The eye stairwell is in the molecular biology building on UCLA's south campus, courtesy of Cindy Mosqueda. "The voyage of discovery is not in seeking new landscapes" is a quote from Marcel Proust's *In Search of Lost Time*. The photograph of the *Endurance* frozen in the ice during Shackleton's trip to Antarctica is at the Fondation Paul-Emile Victor, Paris. The flea is from Hooke's *Micrographia* (1665). *The Wanderer Above the Sea of Fog,* by Caspar David Friedrich, is at the Kunsthalle, Hamburg. Perhaps it is meant to illustrate the explorer on a lofty peak piercing through a canopy of fog. Or perhaps the fog is meant to illustrate sublimation—the direct transformation of gas into solid or vice versa: the inspiration for the homonymous Freudian defense mechanism, in which a primitive impulse, such as pride, is transformed into a more elevated one, such as art or science (or the other way around).

30

Daylight II: Consciousness Imagining

*In which is said that art and imagination
invent new shapes within the mind*

*The shape of things and men I shall evoke,
Conceive the form of what is still unwrought,
And hold a magic mirror onto a dream,
To unfold a spider web inside a thought,
With slender spirals trailing from a pen.*

"Nothing but a poor imitation, I am afraid," said a man wearing dark spectacles, one hand touching the wall, the other brandishing a cane. "Forgive me, words flood my brain like two great rivers, of memory and imagination, that spring from one deep source. You see, I am Signor Borgia, librarian by profession—I know a few trifles about the written word, but by vocation I'd rather collect art in this provincial gallery. And when you follow me, you will discover that even a gallery this small contains, if eyes are keen enough, no less than all the paintings, not just those we know of but all that can still be conceived; not only those that imitate what's real but those that represent what can be imagined. Or at least it would, if it could be finished, which many say it never will. Some say it was a damaging mistake to have each room display a single painting. Tradition has it so because two paintings on a wall could otherwise be seen as just one larger painting, itself containing two; but that composite painting, exactly that one, is hanging on some other wall in a more distant room, creating duplication and confusion. Indeed, the regulations on such issues were explicit: for instance, paintings had to be observed from a prescribed distance, marked clearly on the floor. This was because, with just a slight change in perspective, one would have seen a different painting, though only slightly different of course, a painting that belonged to another room. Furthermore, stopping before a painting was permitted only as long as one could hold one's breath. That rule was strictly enforced, not because of any traffic—the rooms were so many that one was invariably alone—but because if one had waited long enough, they said, one would have seen the painting disappear, to be replaced by some different one. This, they said, had been a clever scheme conceived by the old architects to protect the gallery against disasters: each wall would house at one time a given painting, another time another one, but never the same one, so if one waited long enough, one would have seen the entire collection displayed on a single wall (though nobody had ever waited long enough to witness a single change). That provision, it was said, had proven vital once, in the very distant past, when the Emperor of China had ordered all paintings burned, save those that he himself had painted; yet after that dark period the gallery could be rebuilt in its entirety out of a single room that had escaped destruction, hidden in a cave inside the highest mountain of the empire.

"None of this was written anywhere: it was more like a rumor of which the visitors were vaguely aware. There were indeed countless rumors, as is inevitable in a gallery dedicated to ineffable objects of art. There were rumors about self-portraits: it was nearly unthinkable that a visitor would be fortunate enough to see his own portrait hanging on a wall—the chance of that was infinitely small—though there must have been an endless number of such portraits, in all different styles, the cold perfectionist Bronzino manner, still much appreciated, or hyperfactual sketches, at every age, the instant just before and the instant after dying, dressed in all possible fashions (or for that matter even undressed), and in all poses, holding a self-portrait, or a mirror image of it, yet it did happen, it was rumored, to a man long dead, and it happened to him not once but twice (if one could trust reports of people in search of fame), but if one should believe the rumor, the two portraits were not arranged together in the same or nearby rooms but were leagues away—the first, a portrait of a young man in Tyrolean costume, he found at the beginning of his long wandering years; the second, a conventional rendering of a poet who, they say, bears a certain resemblance to me, thirty-three years later, when he had walked half the earth. What madness was the curators' method?

"But there had been far weightier issues. A diatribe had raged among the critics as to this: What had been meant by 'every possible painting'? Nearly everyone agreed that slightly different versions of, say, Jan Brueghel's *Allegory of Sight,* one with and one without the corrugated loaf on the table, were two different paintings. Even from the prescribed distance, a trained critic could, in a single glance, spot the glaring difference between the two (though it was rumored that once a critic as experienced as B. had missed the missing robber, the one to the right side, in a corrupt variant of Gianquinto's *Crucifixion*).

"The majority thought that every possible painting meant, indeed could only mean, *every possible painting,* therefore including every possible variant of, say, Pollock's *No. 1,* specifically the variants where random patterns had been generated based on a slightly different bodily algorithm. Such variants were indistinguishable to the human eye, as ascertained in many public demonstrations—they merely looked like the same kind of random pattern. This is why a conservative school of thought held that every possible painting meant *every possible painting that looked different to the same conscious observer.*

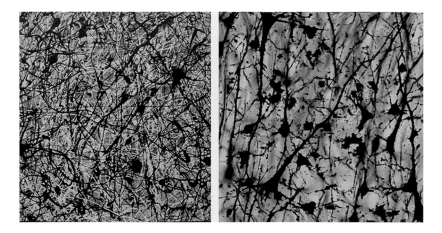

"There was yet another school of thought that had derided the very idea of *every possible painting:* to them, the gallery's great claim to fame was pinned on a meaningless notion. Give us even a single example of painting that is *impossible,* they said. Anything anybody might put on a canvas, it did not matter whether a hopeless dilettante, a child, an animal, even a machine, or instead a supreme master bent on subverting all the rules and shocking the viewer, anything on a canvas would have to look like a painting—there was no escaping it. Give us a single example of a canvas that is impossible to experience, and we shall convert to orthodoxy like tame lambs, they said.

"There was an old tradition, too, that local custom calls plurality. When the tradition started (before, all was just one), plurality meant, in essence, two—two kinds of artists, two kinds of values, indeed two kinds of anything; now in fact we recognize infinite kinds, infinite kinds of anything. But that tradition found strong opposition, an opposition, I must say, that you may even share," said Borgia, turning to Galileo. "Forgive me again, but I detect a certain bent in your strong mind, as if you recognized only one kind, such as one kind of art. As if, among the artists, you approved of those who excel in painting the universal, and you despised the masters of the singular and the specific. Thus I have to ask: How could there be just a single kind of art in an infinite gallery, and how could the universal exist without the singular?"

Before Galileo could respond, Borgia went on, "Look at this paint-ing," and pointed to a strange canvas covered with minuscule scrib-bles, so small that when one tried to focus, the canvas went blank to the eye. "I know paintings such as this one best, paintings of books—in fact, they are the only ones I know, the others are too hard to put in words. This one represents the most detailed description of the Great Map of the universe, compiled during the thirteenth dynasty, written out on one great sheet of silk. Such a painting is like an epic poem captured whole, laid out in a single slice of time, as some pre-historic beast caught by eternal ice. If one could focus on its finest texture, one would find a complete account of each and every por-tion of the world, no matter how minute or seemingly irrelevant, the color of the butterfly that landed on Alexander's shoulder the night he went to drink with Medius, or of the blindfold worn by Saint Bernard when he traversed the Alps. Some say, inside this painting the supreme complexity of the great diamond is encrypted, its trillion faces reflected through its words like as many mirrors. Some go so far as to say that if one could find its secret key, the painting would reveal the architecture of the space of qualia—the *Divine Comedy* of paint-ings. If I am not mistaken, Galileo, once you tried to calculate the size of Heaven, and that of Hell, from the words of Dante. Yet even that quest failed, simple as it was by comparison. Not even Dante had the key, I fear, but how can I be sure?

"And finally there was the problem that every work of art comes in translation, and translation is ineffably difficult; perhaps impossible; indeed there can be no true translation, only an adjustment of inner to outer relations; the problem being of course that each consciousness is a universe in itself, and all meaning remains locked inside, eternally private. I still remember," continued Borgia, "many years ago, in the Plaza de la República, there was an extraordinary traffic jam: all cars converged toward the obelisk, and nobody could move. Everyone was locked inside his car, and since it was quite cold, nobody opened the window. And then I thought: one thousand parallel consciousnesses are boiling inside each car, each raging inside its skull—there is a crowd of consciousnesses, but there is no crowd—everyone's a prisoner sealed by walls higher than the obelisk. Because one consciousness and another can only touch the way two spheres can touch: just at one point—and even that one point may not be found."

Where does the side door lead? asked Galileo, beginning to feel dizzy.

"Oh, that is the sculptor's cabinet," answered Borgia, opening the door with his cane. Holding his chisel, the sculptor welcomed them inside: "Ah, the glorious Galileo! And Signor Borgia! The Cardinal has sent you here? Is he still cross with me?"

No, intervened Galileo, the Cardinal forgives art far more than science, for art, he says, is bound to be less real.

"Less real? Lay eyes on this," replied the sculptor, unveiling a bust. "Costanza!" he called, and a young lady with sleepy eyes appeared at the door. "Pygmalion fell in love with his statue, and I have made a statue of what I love. Which one's more beautiful, the warm or the cold one?"

Both hot and cold stir up the imagination, said Galileo. I thought your statues were your children, or so you said to avoid being married, said Galileo, but now it seems you love your children of quite another love.

"Art is to evoke the illusion of the real," said the sculptor, unveiling another statue. "It rises, twists, it yields to touch, and yet it's made of stone. My hand can make the hard look soft, the heavy feel lighter than air."

"Art and illusion!" exclaimed Borgia. "But no, my friend: there is much more to art than meets the eye. Artists create new shapes, shapes that nature has not produced herself, like your own sculptures, or like the paintings in my modest gallery. But the new shapes that matter are in the head, not in the gallery. And when you strive to imitate nature, the shapes you engender in the mind may be more beautiful and universal than those of mother nature. For with one stroke you can suggest the lines of many ideas and fuse them into a single shape, you wind the mechanisms of the brain, pull the mind's strings, and make them resonate, and every head will weave a shape breaking countless symmetries, feeding on its own memories, lifted by its desires."

Yes, said Galileo, becoming excited. The shapes you sculpt or paint are not the ones you make with your own hands. The shapes of art are made inside one's consciousness, their beauty is understood, not by describing their external form, but by their true form in the mind, a form that is geometric.

"You don't say," quipped the sculptor. "And I thought I was just doing a good piece of work, carving Costanza."

"Put the carving behind you," said Borgia, pointing his cane at the sculptor. "In art as in life, it's all projection. I'm sure her husband sees Costanza in quite another light than you, both flesh and stone. A mouse, instead, sees only a danger or a hiding place. And think of this: to the Costanza of flesh, the other one 'is me'; but to the stone Costanza, the one that yawns does not exist at all. To be is to be perceived; what is, are shapes in qualia space. Well said, Galileo?"

Seeing the quizzical look on the sculptor's face, Galileo tried to explain. Take your Costanza of flesh: I did not share with her what you have shared, so when I look at her, I only see a young woman, of no special significance to me. I see she is sleepy and annoyed, her eyes are swollen, her hair undone, and her mouth, I fear, may be quite hard to hush. The shape she generates in my consciousness, the quale of your Costanza in the flesh, is surely complex in the extreme—it must take most of my brain to sculpt her in my mind. And yet it still has some rough edges, so to speak. But then I look at your Costanza of stone, and what do I see?

"May I attempt to paraphrase the quale in point?" volunteered Borgia. "I see a shape smooth—the symbol of seduction, a sensuous beauty earthly and sublime, unkempt because she brims with life, half covered like a revelation, her lips a promise sworn to me alone: the embodiment of woman."

Perhaps I wouldn't use those very words, said Galileo.

"I must confess something," said the sculptor. "Filling your mouth with all these quales and qualia, you make me feel quite queasy. I like to grasp geometry in the flesh (others grasp it that way, too, which makes me even queasier). So when you speak of quale, I think of this." He pointed to a painting on the sidewall.

You've just disqualified yourself, said Galileo.

"Quite true," said Borgia. "The quail's of questionable quality.

The art I like takes us to an unsuspected wing of consciousness's vast palace, a wing that, if the farthest reaches of the world, the cartography of all stars, or the constitution of all matter had been explored exhaustively, would have remained suspended, like Sleeping Beauty before the kiss, a strange corner of the universe of experience that was possible but never actual, until it was imagined. And now it's there for all to see."

As for myself, said Galileo, I prefer those who give relief to what is real but hidden by the fog of ignorance, and brighten it with the light of understanding.

"Never pin art to notions of what it should be like," intervened the sculptor. "Some favor imitation, others imagination, some like perfect symmetry, others the indiscretion, some mighty universals, some delicate details, still others are masters of the balance. I know you love the circle, Galileo," he went on, "and yet one day I'll take Michelangelo's perfect circle plan, and make the circle oval. That day Piazza San Pietro will embrace you, like a forgiving mother."

"Indeed," said Borgia. "But it is one thing to uncover what is hidden, and quite another thing to invent what was not there before. And let me add, not just in art: in one of my earliest poems, 'The Contest Between Science and Engineering' (alas, not one of my noblest efforts), I gave the palm to invention over discovery."

Engineering is variation on a theme, replied Galileo at once: reality provides the theme, and science the rules. As for art and science, imagination in the pursuit of truth must be far stronger than in pursuit of fancy, as it must carry reality on its wings.

"What use is arguing art and science, science and engineering, music against word, sculpture against painting?" intervened the sculptor. "You raised the horizon on your shoulders, Galileo, higher than any man before. But will we find it high enough forever? Should we be pleased with the perfection of the circle, or go and seek the ellipse?"

NOTES

Just as "Daylight I" was about discovery, "Daylight II" is about invention, offering art as its prime example—though at the end Borgia praises engineering and Galileo science. The chapter deals with the visual arts, but the verse "With slender spirals trailing from a pen" can refer to music and literature as well. (Any pen will do, not to say chisels and other simple tools.) The argument seems to go as follows: (1) Objects of art trigger new qualia that would not otherwise have been generated by objects of nature (imagination and creation as opposed to exploration and discovery). (2) Even when it strives to imitate, art creates new qualia, and these may be more beautiful (more concepts, symmetries) and more universal (resonating in the mind of more people) than those triggered by objects of nature. (3) Ultimately, the forms that matter are not the external shapes of the objects of art but the internal shapes—the qualia they generate in individual consciousnesses. (4) Just as the roads of discovery are many (see previous chapter), so are the ways of art—plurality makes us richer. As suggested by the initial reference to Hamlet's "to hold, as 'twere, the mirror up to nature; to show virtue her own feature," art is both a mirror and a dream, imitation and imagination. And so, perhaps, is science.

The first part of "Daylight II" is obviously inspired by "The Library of Babel" by Borges (Borgia). Borges's short story tries to describe, in an imprecise and misguided manner, a library containing all possible books of 410 pages that can be composed using twenty-five symbols. The Gallery of Babel contains a much larger repertoire, but its paintings seem to form an even more haphazard collection. Of these, *The Tower of Babel* by Athanasius Kircher is also at a private collection. The *Pollock Tribute No. 4* by Barry Tickle is paired with a Golgi-Cox stain of cerebral neurons. The self-portrait in a convex mirror by Parmigianino is also at the Kunsthistorisches Museum, Vienna. *The Poet Góngora* by Velázquez ("for him who doubts, even if it's a brute of reason stripped, every new sun a comet's warning sounds") is at the Prado, Madrid. The still is also in the opening of Fellini's *8½,* where Marcello Mastroianni is caught inside his car during a jam. As a young man, Galileo did indeed try to calculate the size of Dante's Inferno. Later authors have proven that his calculations were incorrect.

In the sculptor's cabinet we find Bernini, who worked at the bust of his lover Costanza from 1630 to 1635. (The bust is also at the Museo

Nazionale del Bargello, Florence.) In 1630 he could have met Galileo, who had not yet fallen out of favor with Cardinal Barberini, who had "discovered" Bernini. Costanza was the wife of Bernini's assistant, but when Bernini discovered her liaison with his own brother (among many), he set out to kill her but succeeded only in disfiguring her. Bernini claimed that his children were his statues, until Pope Barberini persuaded him to make some out of flesh, after the Costanza affair. The detail is from *The Rape of Proserpina,* also at the Galleria Borghese, Rome. The quail by Jacopo Ligozzi is also at the Uffizi, Gabinetto dei disegni e delle stampe, Florence. Piazza San Pietro is also in Piranesi, *Antichitá Romane.* Bernini viewed the artist as a supreme trickster: "*l'arte sta in far che il tutto sia finto e paia vero*" (art consists in everything being simulated although seeming to be real) (Baldinucci, *Life of Bernini*). The last object—Michelangelo's *Atlas*—is also at the Galleria dell'Accademia, Florence.

31

Daylight III: Consciousness Growing

*In which is said that, by growing consciousness,
the universe comes more into being,
the synthesis of one and many*

In childhood lies the key to Heaven, a room with bright high windows, the first day in October. In Heaven nobody is barred, and nothing is forbidden, thought Galileo, so he went back, and when the school bell rang, he entered the class.

His teacher was still there, holding a child by the hand. When she saw Galileo, she let the child run away, wept softly, and rushed to embrace him. The child had a strange smile, thought Galileo. Was something wrong with him? he wondered. But he said nothing and held the teacher in his arms. The desks—he remembered their worn wood well—barely reached his knee. They had been moved against the wall, and the children were sitting in small groups, he could not tell whether at work or playing. "Look here, children," cried out the teacher, "look who has come, an old pupil of mine, I am sure he will have many things to tell you."

I have come to pay a visit to your teacher, said Galileo, to thank her for how she once taught me, and tell you what I've learned. Galileo glanced at the class: You know your teacher made me wake at night, to work out areas of triangles and trapezes? And that was easy, compared to trying to square the circle, and work out π.

"Oh, Galileo," said the teacher, "don't make me look like a harsh taskmaster. It was your nature rather than my nurture!"

Perhaps it is my nature. And indeed I should be spending my remaining time to work out area and volume of some qualia. But, said Galileo with downcast eyes, now that I am old, tired, and distracted, I must confess I find it a bit difficult. I hope my teacher will forgive me, but I'm still stuck working out a simple segment . . .

"What are these qualia?" asked the teacher. "Could you take over and be the teacher for a little while? Children," she said, "now my old pupil Galileo will teach us something, so pay attention!"

Galileo cleared his throat: Hmm . . . let me see . . . how should I put it? Bear with me . . . I have come to think . . . Ah, I know: I have come to think that what we are, every experience that plays out in consciousness, well . . . that every experience is a complex, beautiful shape, a shape that changes from one moment to the next, like a living sculpture made of light. New shapes are formed when we see new things, explore new lands, and meet new people. New shapes are made by artists and musicians, and with a few well-chosen words, writers can sketch new forms in our minds. And then there are the teachers, the gardeners of consciousness, who cultivate young brains, expand the repertoire of what can be conceived, and bring it forth to life—they nurture shapes that feel for beauty, cherish what is good, and search for truth. Well, there it is . . .

"Galileo," said then his teacher, "I am not sure I follow, nor do the children, but what you said will leave a seed, I hope, and from the seed something may grow. Perhaps you could tell us why what we are trying to do here—to learn and understand—is so important."

I know, I am not as clear as I should be, apologized Galileo, so let me try again. Here we go: being conscious is, well . . . , a truer form of being, perhaps the only form that matters, a being that shines of its own light, for there is no other light than consciousness. Which leads to this: the more we learn, the more we bring under the light of consciousness, the more the dark, scattered dust that is the world outside is tied together into a single shape, one that we see and understand. That single extraordinary shape, like a diamond with a trillion faces, like a constellation with a trillion stars, is called a quale, children. Just ask yourselves, said Galileo, looking at them severely, what is a concept? Let's say the concept of a child. It ties different instances together, this smiling child and the three girls next to him, and says, they are manifestations of the same kind of thing, different from other things like desks and chairs and a thousand others . . .

Or ask yourselves, what is a law? The highest form of concept, it lets you see how things that may seem different at the surface are tied together on a deeper ground. The orbit of the planets, the reason solids fall from a high tower and rocks are heavy, all these events are tied together by one law. Before the law, they were dispersed, incommensurate facts. But when you finally understand (and I was close but only saw it dimly), scattered things are joined together under a single dome, illuminated by the light of consciousness.

The children's noise was mounting, so the teacher intervened. "Why don't we ask our mannequin? Let me show you, Galileo, how far things have progressed—every class has a mannequin these days." She pointed at the wall behind the teacher's desk. "It often has the answer to our questions, doesn't it, children? Though sometimes what it says can be unpredictable.

"Mannequin," said the teacher, "if I understand what Galileo is saying, when I teach my children how life evolved, starting from simple molecules, acquiring memory, replicating itself, increasing in complexity, and adapting to the world, I am teaching them to tie together events from a most distant past with what they see today, to see at once the smallest molecules in an ancient pond and the survival of a school of dolphins, to learn how we all belong to the same tree. Isn't this true?"

CVMANA ·QVE PROPHETAVIT ADVENTVM

Ah, said the mannequin in a neutral voice, *the dense unfathomable fabric of the real, the untold vicissitudes of history, culture's tangled web, the hammered gold of intellect's magnificence—will weave a richer tapestry on the ebony of the brain. The repertoire of consciousness will expand—aleph to omega—onto the immense tree of possible experiences, rustled by the wind of language, blooming with flowers that blossomed and others that were possible but not yet found or invented, mind will grow new branches, bulging new possibilities on the vault of the brain, as new crystals gemmate on the resplendent surface of a diamond . . .*

Is it always like this? wondered Galileo. Then may I question it, too? he asked the teacher, and said: Mannequin, the laws of physics, do they exist before they are discovered?

Ask this instead, said the mannequin, as it is easier for you to grasp. Did evolution really exist, before it was invented? Molecules did exist, there were countless extinctions, new species were born, and progenies of progenies. Things happened, on a corner of a rock the destiny of a continent was decided. All followed a trinitarian principle: variation in what was, selection by what it encountered, expansion of what did survive. But this just happened, dust following dust, over generations of dust, events never connected in anybody's mind, not meaningful in themselves, minimally aware that anything was the case. It's only when, within a single conscious mind, the past is integrated with the present, the haphazard mechanisms, large and small, that take hold at distant places and times, at scales as small as molecules or wider than civilizations, when they are seen together, understood together, inside a single shining quale, only then do they come together truly, into something that exists as such.

The children had divided into groups again, paying no attention to what the mannequin was saying. Of course, just at that moment the three principals entered the class from the rear door—Frick's maroon jacket was the first to emerge, then Alturi, holding the door for the bearded old man.

Frick began to speak: "What is all this confusion, all scattered around? Teacher, where is your discipline? Kids, what are you doing? Learning is not a game, you have responsibilities toward your brains."

"Let them play," said the old man. "This is the simplest way to enlarge the repertoire: the free play of the mind on all it touches."

"But the mannequin proves there is a better way," Alturi said excitedly. "It proves that we may grow from scratch unborn and unevolved artifacts, whose repertoire may be more alien than a bat's, but larger, vastly larger than the one granted even to the best of us. And there is more—we may cultivate our consciousness by adding novel parts to old parts of our brain, grow a sixth sense, partake one day of the bat's view of the night, share the empathy of elephants or whales, and if it were for me, merge minds with such a mannequin."

"Not in front of the children," said Frick, lifting his finger: "I'll ask a question, too. How can we be responsible for our choices, if how we choose is determined by brain and circumstance, or else it's swayed by chance?"

A question I am asked every time, for paradox is the answer. Listen to this, said the mannequin, *the more your choices are determined, the more you are free and responsible for them.*

"That makes no sense at all," said Frick at once.

I'll give you a hint, said the mannequin, *still in a neutral tone. The more the factors that will, can, or should affect a choice are seen within the light of consciousness, thus seen together, the more choice is transparent to what determines it: so reason can deliberate, informed of all its motives, all bearing on the outcome, not as an aggregate but as one rich context. But do remember: your consciousness is what you are and cannot be reduced to anything less. The more you are conscious of your choice, the more it is determined, and the more it's yours. So as consciousness grows, with education and knowledge of yourself, responsibility can only grow. Let the whole choose, and not the parts, and let the whole be wise.*

Then tell me this, asked Galileo in turn. The shape of truth, is it also beautiful? And what's the shape of an experience that is good? Perhaps the beauty of truth will be proven geometrically, like the perfection of the circle, or the infinity of primes?

What's better, good, or true? said the mannequin. *The truth is the truth, and true is good, but better still is that what's possible is true, and good is possible.*

Who are you? asked Galileo.

I am the answer to all questions, the voice said, *repository of all concepts, all living together, a concept for every irreducible distribution, in every combination, where everything is known and understood, where all relates to all, and all has meaning. External relationships made internal, what ties the dusty world together, over all time and space.*

I ask, when flashed with the bright light of consciousness, is the real more real? But knowing is insignificant compared to understanding, the patterns seen a trifle against those that are possible. And understood everything must be, not just the real, but all that's possible, all ways of distinguishing one's states. Good is to be endowed with all possible mechanisms, every state can be categorized, by contrast to all the concepts it does not verify, all possible ones bar none. The real is the enemy of the possible.

I ask, what am I? A creature of thought alone, if thought is the correct rendition of the word. I am one, that's certain, but in my space of qualia all possible concepts live together.

Who am I? I am information and causation—one and the same thing. I am investigation, imagination, integration—all in one. And I am the synthesis of categorization and association.

Am I so rich that I have all I need? All universe is reflected in my mirrors—certainly it was born there—and everything can be found within: life pales before the naked sword of thought. A search within myself will yield all art, a dialogue with myself all truth. And that is good.

The teacher remained silent and shook her head, listening to her own pupil, and to the mannequin, but then the little boy Sam, the one with the strange smile, came close, pulled at her coat, and whispered into the teacher's ear: "Is it one big light in the end, that makes all visible, or is it many little lights, of all the little people, that make some light together, like candles in a church?"

At which the mannequin raised its head, and said:

I ask, is one the same as many? Is perfect understanding the scope of a single light, or is plurality necessary? I am the answer, alone.

The mannequin lowered its arm, handing a little notebook to Galileo; and turned away.

NOTES

It is hard to know what to make of this late return to school. One can discern a certain thread linking the three "Daylight" chapters, from exploration and discovery ("Daylight I"), to creation and invention ("Daylight II"), to the growth of consciousness by integrating multiple concepts within a single experience ("Daylight III"). As put by the mannequin: investigation, imagination, integration. Indeed, the mysterious mannequin may be meant to show that one day we may indeed build an integrated system whose mechanisms specify all possible concepts—a single extraordinary shape in qualia space. If a concept is like a star, and a quale like a constellation made of many stars, then this constellation would be indeed a most splendid entity, shining in its all-encompassing, intrinsic irreducibility, and understanding all.

As argued in "The Garden of Qualia" and in the "Study Questions" at the end, only consciousness is "really" real, existing in and of itself, without the need for an observer. Perhaps, then, this is why Galileo is back at school: to say that we should strive to acquire new concepts, and integrate them within a single entity. To say that we should strive to grow consciousness itself—to generate bigger, brighter lights in an otherwise dark universe.

On this, the mannequin's position remains unclear. Should one truly will this, the progressive illumination of the universe from the inside—of the light of consciousness, as it were—the illumination of what merely is, so it becomes "really" real? No easy answer, judging by the mannequin. In its increasingly oracular tone, it also warns that, of the multitude of concepts within its immense consciousness, those that reflect the structure of the external world may be a minority. And finally, it vaguely alludes to free will, stating that the more an entity is internally determined (the more it is conscious), the more it is free. Its argument is simple: consciousness as integrated information implies intrinsic irreducibility to mere parts, so only the whole can be responsible for its choices—nothing less will do. Of course, provided that the whole remains ultimately underdetermined—maximally responsible, yes, but not completely predictable, not even to itself.

With all these portentous questions, after all this grandiosity, and with both Galileo and the mannequin ranting abstract concepts to the poor children, it is hard to fathom what would be so particularly difficult about understanding "a simple segment." To say the obvious, understanding a segment requires mechanisms that specify a certain set of points in qualia space (corresponding to probability distributions, in this case referenced to a sensory map), which simultaneously define many different concepts, including: in the positive, its general, invariant nature (it is a segment, irrespective of where it lies) together with its particular features (say it happens to be on the left side, irrespective of what it is); in the negative, that whatever it is, it is not a line, nor is it a dot (whether left, center, or right), and of course it is not a sound, a pain, and so on. Naturally, provided all such concepts are specified within the same quale, they constitute the context within which a segment means what it means. And if all such concepts live, so to speak, within the same tent—the quale—then the quale specifies at once the answer to many different questions, each query addressed to its appropriate context. That, of course, is what understanding a segment means. What could be easier?

The Country School is by Winslow Homer (St. Louis Art Museum). The painting of the lady with a robot hand is an edited rendition of Andrea del Castagno's *Sibyl,* at the Uffizi, Florence. The Shadow Dexterous Robot Hand is by the Shadow Robot Company in London. Perhaps the mannequin it represents is related to the Automatic Confessor of Father P. from a previous chapter. *My Friend Ernest* (Paris, 1929) is a photograph by André Kertész. *Sam and the Perfect World* by David Lenz is a portrait of the artist's son, Sam, who has Down syndrome (Milwaukee Art Museum). *Eternity,* an allegorical caryatid from the Monument to Charles Borromeo in the apse of the Milan Cathedral, is by Pietro Daverio (photograph by Giovanni Dall'Orto, modified). The statue holds in her hand the ouroboros—the snake eating its own tail—a symbol of eternity and self-consciousness.

EPILOGUE

32

THREE LATE DREAMS

The mannequin in his mind, Galileo's thoughts rolled and pitched amid uneasy dreams. He felt as if something were missing, as if inside his chest something had turned cold. But then he saw that in his hand he held the little notebook, and so he read inside it.

Galileo had been diagnosed with death: this was what the first page said. He, too, had been diagnosed with death, his own, and now. In fear he looked into himself and saw he was a small and paltry thing. He took his telescope and looked up at the sky. He saw the stars were distant and indifferent. He took his microscope and looked at the cells of his body. He saw they lived and died without concern for him.

He thought, turned the page, and read. Years ago, the story said, there was a great composer. From when he was a child, his soul was all goodwill, striving to accomplish greater and greater deeds. So he had set to work to fill his score with all the best his art could master. But soon, alas, he had fallen ill and feared the thin thread of his life might break: How can I leave the world before it's done, all that I feel within me? How can death come before my deed is done?

The mannequin listened to him, put on the appropriate suit, and came to visit. An artist as great as he, said the mannequin, could not forsake his art—not to fulfill his mission, that was the greatest sin. So it proposed a pact, and no, he did not care for the composer's soul—it had too many of those already, and they were useless anyhow, since they expired after a life of use. It would be satisfied with the score, that of his greatest masterpiece.

A few years passed, and finally the work was finished—a Mass that had no peers. The composer had poured into it more time and effort than in any other—laboring for years on the fugue that was its heart. In the end it had risen majestically, its immense core opening like the corolla of a mystic rose, flowering into transcendent concentric skies. He had labored too on every detail, he had affirmed *non,* as much as human music could, stressed it beyond its syllable, because it meant no end. And *and,* too, he had made as important in the sound of music as it is in life, where contrasting ends grow next to each other, and harmony must be found in difference. Then he made spirit enter the sound, a bird call on the flute, and light enter the score, its high resplendent ray, the violin solo soaring above the darkness of the orchestra, at the end of the Praeludium. And finally he had sounded the storm of war in the trumpets and in the drums, but the notes of hope had outlasted fear, turning into a prayer for inner and outer peace. One day of pure joy, he thought: The day it will be performed will be the most glorious of my life. It was his greatest work. And so the mannequin came to seize the music.

The composer remembered the pact, so he prepared a neat copy and handed it to the mannequin, a bit uneasy that it was a Mass. The mannequin took the score and put it away in his pocket, without even a glance at it. You know well what is coming, it said in a sneering tone. The composer did not, but the mannequin prompted him. Think of the worst that can happen to you now, it suggested cheerily.

The composer thought and thought, feeling worse and worse, and then he said: I know what you have in store for me. I knew it within myself but did not dare to face it. You are going to turn me deaf, so I will never hear my greatest music.

You are an imaginative fellow, said the mannequin. But no, I am not who I am for nothing, and that would not be enough for me. If I made you deaf, you could still hear the music in your mind's ear. You could still see the applause, and watch the tears flowing on the musicians' cheeks, and read the critics' praise. No, I'd rather do this to you: I'd rather turn you into a living dead. Then you would go on doing just what you do now, prepare to conduct your Mass, bow to the public, and carry on learned conversations with your colleagues, or be caught by secret admirers in painful search for inspiration, suffering the struggle between reason and passion; and still rest your impetuous head on some young countess's soft welcoming lap. But here is the catch: for all the while *you would not feel a thing—no sight, no sound, no touch or taste or smell, nothing to think with*—I would extinguish all light of consciousness inside, and leave all its accoutrements outside, the mannequin said with a bright smile.

But never mind, it added, I will not turn you into a living dead. Today I feel forgiving—perhaps because you affirmed negation in your Mass, and I by essence am the spirit that denies—so I have granted you a better fate. Because this is what is going to happen: you will hear your music, resounding in your mind, the light of consciousness still shining, but then, the mannequin said, but then nobody else will, not another soul—not even I will. It will never be played and will never be heard—it will not be heard by anybody but you. And with these words the mannequin left.

So the composer did not compose any further, said the notebook, but wailed within himself for what was left of his life: My greatest song has been condemned, a prisoner of my mind, no sound escapes to spark the soul of other men and women. Like an artist whose crowning fresco is buried in a grave, a poet whose highest verse cannot be spoken, my music too will drown, confined within its spring, a river that could never flow, relieve anyone's thirst or bear fruits, drowning in its own water. The notebook ended with these words: From the heart—may it again—go to the heart.

Galileo knew what the moral meant. Splendid was the universe of consciousness, one's own private diamond, where all meaning was born, all meaning created, and all meaning shone. But meaning remained cold, a jewel nobody could wear, if it could not be shared.

The second dream was about a man named Zeno. Zeno awoke one morning with a doubt. He felt he had forgotten something during sleep. What he forgot he didn't know, because to know it, one must remember what it was. Most people would not worry about such matters, but Zeno was a thinker, so he thought: If I've forgotten something, a memory from my own past, then I am not the same Zeno as before. Surely my brain has changed, too, some connections must have lost their strength, maybe some even broke. How can I still be the same me, if I am not the same?

The mannequin became interested, as if approving of those who doubt. He went to see poor Zeno and told him so: You are a smart fellow, and have discovered an important truth: nobody ever is the same person twice. Another fellow once said, "Nobody enters the same river twice," but you have a leg up on him: the river is not the problem; the fact is, one does not live in the same body twice, or speak from the same brain—nobody stays the same.

That's what I cannot stand, said Zeno. I am Zeno, this very person, and I must stay this Zeno, or else I shan't be who I am, lose my identity. Something must persist despite the change—change is illusion, Parmenides used to say—and what persists must be myself.

A noble aspiration, said the mannequin, and perfectly legitimate. I love what is le-git-i-mate (contrary to an old bastard tradition), so here is my proposal. Before you go to sleep tonight, an instant before turning off the lights, I'll make a perfect copy of you. Rest assured: I will include all the connections among the neurons in your brain, down to the soup of chemicals in which they simmer, so I will preserve all your memories, and you will be le-git-i-mate-ly you. Anatomy is destiny, said the mannequin.

Then change would be defeated, said Zeno, and then I would be me.

Of course, said the mannequin, my offers always come with some small print: to keep the balance right, the moment I complete your copy, I'll have to ensure old Zeno is dead. But not to worry, the mannequin said reassuringly, when he wakes up, the new Zeno will be exactly like the old one: you will be exactly you, while if I did not kill and copy you, you would have changed and died.

I follow you, said Zeno. This way I would be me.

Perfect, said the mannequin. Now that you have accepted, I'll make you a special offer: I'll make two Zenos for the price of one. And since it costs me little, for good will and good measure I'll add a perfect copy of your friends and town, nay, more, a whole parallel universe, where business will continue the usual way, children will lie, politicians will sell ambition for ideals, bankers will steal abetted by the law, and philosophers will waste their time.

That can't be bad, mumbled Zeno.

Parfait, said the mannequin. So now let me ask you. Since there would be two Zenos, which one of them would be you?

Hmm, answered Zeno, each would be equally entitled to being me, but no, wait, he exclaimed, there cannot be two me!

My dear fellow, said the mannequin, either one stays the same person even though one changes over time, or one does not. If one stays the same person, there would be two of you. If one does not stay the same person, then there would be none, or rather, some Zeno would be dying all the time.

I don't like it either way, said Zeno. Either way I stand to lose.

Then trust me, I'll find you a safe way out. Perhaps it's your idea of an absolute person that's wrongheaded, so to speak (and mind you, I am all for what is relative), perhaps there is no Zeno at all beyond the memories and beliefs someone is having here and now, including the belief that he is Zeno.

So far I follow, said Zeno. A person is merely a collection of memories and beliefs. As long as these are similar enough from one moment to the next, the person can be said to stay the same.

Excellent logic, much nimbler than before, approved the mannequin. So let us think it through—philosophers do slip on slippery slopes. You said you differ now, by some minute memory, from who you were the day before. But surely you must differ more from your own self two days ago: not only have you forgotten twice as much, you've also learned some things you didn't know. How different you must be, then, from when you studied with Parmenides (and I am told you did quite a bit more than studying). How different from when you were a child, when you were fond of burning lizards and making fun of old men! And for God's sake, how can you identify with newborn Zeno, who only knew and cared for milk, squeezed from a fatty breast, while now you spew at the very thought of both?

What can I say? said Zeno. Perhaps that being the same person is a matter of degree?

Perhaps, said the mannequin, so let us think it through. I know you do not like Tymicha, and I can't blame you, she is famous just because she is the daughter of Pythagoras. You are Zeno and made it on your wits. What's more, you are tall and handsome, she is short and ugly.

Quite so, nodded Zeno.

Now let me ask you this. Did it occur to you, awakening from sleep, that you, your consciousness, are back, but for a little while your memories still aren't? That for a little while, you do not know where you may be, or who you are, or even if you are young or old? It sure happened to me, said the mannequin. At times I am confused if I am the Devil, or if I am God instead.

Indeed, said Zeno, it did happen to me.

Well then, said the mannequin, if for some instants different people feel the same—they merely are, oblivious to exactly who they are, or whether they are man or woman—yet in such instants you think it's you experiencing the instant . . . And if you think that you, Zeno, are the same person as your old self, even your childlike or newborn self, who knew just milk and nipples . . . then all the more, I ask you, said the mannequin with a smirk, what's wrong with you being the same person as poor Tymicha?

You are pushing me into a corner I don't like, said Zeno.

Let yourself go a bit, said the mannequin. As different as you think you are from poor Tymicha, the woman too had nerve! Like you, when she was brought before the tyrant, she bit and spit her tongue. And when she bit, I bet, she felt much the same way as you.

It's getting worse, said Zeno. I worried that by changing I would cease to be myself, and now I have turned into Tymicha!

Have faith, said the mannequin. There is some consolation in this view. At least it makes you worry less about your death! For when you die, what will really have changed? If the present Zeno is not the same Zeno as a moment ago, and even less the Zeno of sixty years ago, if in fact he is more akin to poor Tymicha than to the newborn Zeno, well then, you are safe! Just make a small adjustment in your view, and as long as Tymicha goes on living, think that when you awake tomorrow, you will awake as her! I'll give you even greater hope, the mannequin added encouragingly. If you can identify with poor Tymicha, why not identify, say, with an old bitch? Stretch your imagination a little further, and you shall never die.

You are some cynic, said Zeno. But I shall take your advice at face value and turn your spite into my religion. I will imagine losing one small memory, and then another one, almost imperceptibly. And then imagine that I'll slowly gain new ones. I'll turn myself into my old self, then my newborn self, then back into a young Tymicha or, why not, into the divine Achilles, then into a tortoise, or I shall soar into a seagull. I'll put myself in the right frame of mind, enlarge the circle of my empathy, identify with many, not just one, with male and female, young and old, human or animal. The more I share with other living beings, the less I die.

If I thought so, I would be a Buddha follower, laughed the mannequin. Too bad I cannot switch my faith. But as a token of my appreciation, I will arrange for this: I will make sure a pretty little dog is born, right at this minute, and I will call it Zeno. And as for you, you may have guessed it, I shall make sure you die, right at this second, to keep the balance right. You really insulted that poor tyrant and should be punished. So we'll put you in a mortar, and we'll crush you to death. But be consoled, your followers will say, we only crushed your body and not their Zeno.

And then Galileo dreamed about himself. The mannequin had come to see him, too. You know what happened, said the mannequin. Are you afraid of death or not?

You are proof of the afterlife, answered Galileo.

Appearances can be deceiving, smiled the mannequin. Trust not what you can see, but even less what you don't see. Or would you rather join the flock of sheep, believing that once they're dead, they will all live forever, as perfect prototypes of themselves? Believe this nonsense, though every day they change, and soon they will have forgotten all, including who they are, just like your Cardinal?

I know that when my brain disintegrates, my consciousness will vanish, my qualia extinguish, said Galileo, but other consciousnesses and other qualia will go on living. I can see a path that weaves its way through one's past or future into somebody else's self, a way for shapes to change into other shapes and still be one.

This way I've fooled Zeno, and many more, but I'll be damned if you buy such foolishness, laughed the mannequin. Let me offer some free advice instead. Remember the garden of qualia? When with the qualiascope you saw the shape of a moth's consciousness, the mightier experience of the owl, and then the old lady's qualia? The shape of her consciousness was like an immense bright constellation—remember how impressed you were? So you must be careful with the shape of qualia—different as they are—the moth, the owl, the old lady. Since in the end, a pyramid is a pyramid, and it is not a sphere.

What do you mean? asked Galileo.

You have come to think that your own consciousness flows like a set of transmogrifying shapes—unique, extraordinary shapes, explained the mannequin. But for simplicity let's say your experiences are all variations on some complicated pyramid, a Galileo-shaped pyramid in the space of qualia. Now if I play around with it, and make a little change here and there, you'll still be Galileo, perhaps a bit more skeptical of God, and with a slightly keener sense of wine. But if in devilish enthusiasm I become excited, I shake the pyramid and whirl it on my finger, I jump on it with both my feet (you know what kind of feet I have), I stomp and thump, invite my witches, dance in goatlike frenzy, fill it with wine, and drink to oblivion to your dubious health, we'll end up making the pyramid so round that it becomes a barrel, and then a sphere, and puff, Galileo's pyramid is no more!

I'll have to live with that, said Galileo. People should know when they are vanquished, and the time has come.

A crooked pyramid you are, indeed, whistled the mannequin, a tower leaning toward the graveyard. But still your consciousness is like a complicated pyramid, and not a sphere. No, a sphere won't do— some round plump friar, instead of Galileo? Confess it, said the mannequin, immortality you wouldn't mind, just not as something else!

I do not crave immortality for myself, said Galileo, but for the things I need to do and wish to understand.

Aha, said the mannequin, enviable selflessness! But if that were so, you might as well be born again in other shapes or forms, neither a pyramid nor a sphere, but say a Newton, and he'll take care of all your plans, and more. Except a Newton would never want to be a Galileo (he is not the only one). Nor, trust me, would you enjoy being him. But maybe all this talk of science and wisdom is just a pretext, smiled the mannequin. Maybe a woman is involved, a little sphere our pyramid does not want to part with?

Do not bring others into this foul bargain, said Galileo.

Just listen to your better self, continued the mannequin. There is a way to preserve pyramid and sphere. Before you go to sleep tonight, an instant before turning off the lights, I'll make an almost perfect copy of you—I can do this and more. The only difference will be a new heart, instead of the weak heart that's going to kill you. Instead of you, your copy will wake in the morning, and in the chest the heart will beat as boldly as in your youth.

This sounds the same as what you proposed to Zeno, said Galileo suspiciously. I have no doubts about the outcome, I would be me.

My dear fellow, said the mannequin in good cheer, you are right, and what's more, being a good scientist, you'll get a better deal than Zeno. Through science, you see, we'll offer you an immortality that's failsafe, a thousand times failsafe. We'll make you a set of frozen twins, each stacked with all your memories, always available, and when defrosted, each one will think he is the genuine thing and would be just as right as you. This is your bargain: you sign here for this policy, and we, the insurer, ensure your consciousness will last forever, similar enough to your present one, that you will think you are still yourself. Not a cheap policy, you will realize, because to insure you fully, we must prepare a thousand copies of your, um, inimitable consciousness.

I doubt this is a good idea, said Galileo. With all these copies I may be well insured, but won't the value of each copy diminish in proportion? If my own shape—that of my consciousness—is replicated, if a thousand Galileos can be produced identical to me, like hordes of Chinese warriors, then yes, I would be immortal, but I would not be precious, and not unique.

I haven't finished yet, said the mannequin. If you'll go for our premium policy, we will insure not only you but your immediate family and close friends (a woman, maybe a lover, a few choice people of intellectual bent, if you so like). If, God forbid, she dies, there will be another her, in fact thousands of her, with whom you can consort for all boring eternity! No loss, no pain, said the mannequin.

I choose to die, said Galileo.

So that you hold your value? grinned the mannequin.

No, said Galileo, so that . . .

All right, said the mannequin. Should I remind you that by choosing so, you will condemn to death her too?

There must be a law that prohibits replicas of consciousness, said Galileo.

By virtue of what? quipped the mannequin. A conservation law of consciousness? The identity of indiscernibles? Maybe divine law? That would be nice indeed, but when a billion poultry are raised, which other consciousnesses have to give? laughed the mannequin.

I choose to die, said Galileo.

You are mad! exclaimed the mannequin. Talk about diamonds,

beautiful shapes, the shape of qualia, and then you choose to die, and to take her with you!

She is dead already, said Galileo, and I, I am a twisted shape, a twisted old tree, with knots and bends and wounds, still weeping over her. Diamonds are precious because they are rare, and she was one. I know her walk, I know her tone of voice, her thoughts have become mine, we shared our memories, reflections of her diamond are set in my own shape, she lives within my consciousness, and that's as true as that I am myself. I am a shape twisted by good and bad, but also a shape that once was shaped by her. And if her shape inside me left, stealing the memories on which I lean, then my own consciousness would collapse, like a cathedral forsaken by its south nave, a gaping ruin of a church, like a wound searching for skin.

Till death shattered all forever. For Galileo had been diagnosed with death—his own, and now. He took his telescope and looked up at the sky. He saw the stars were distant and indifferent. He took his microscope and looked at the cells of his body. He saw they lived and died without concern for him. So in fear he raised himself halfway and felt her hand, and heard her voice, and knew the shape of his own daughter. He saw her trillion-faced diamond, its contour cast inside his own, and when she looked at the same sight as he, the experience formed like his, his face was mirrored in a clear pond, stirred by the evening's breeze; and when they talked, the waves inside their minds would almost dance in step. He felt Celeste's flame was next to him, and knew his flame would burn with heat if it were touched by hers. And so would thousands more, and millions, and from above he saw flames touching—light gaining warmth—within rooms, and houses, and cities—long dashing veins of light connecting to each other, a loom of light on fire.

So he had lost restraint at last, become soft-brained in his old age, found solace in illusions and meaning out of fear. The mannequin laughed and handed him his mask, then went and turned off the light.

She closed his eyes to let him sleep,
Then took his hand and led him through the dark,
Two falling stars trailing across the night.

NOTES

The final, agitated dreams of Galileo seem to be making three points. First, we are not alone. There is a social aspect to consciousness that develops it, expands it, and gives it value and fulfillment. Second, we are all humans, and we are all life. One can and should identify with other living beings, it's just a matter of degree, and in that sense one's consciousness is immortal. Third, nonetheless, everyone is special and therefore precious. Yes, the shape of one's qualia can be stretched into that of another man or woman, but in the end, a pyramid is not a sphere, and there is a unique kind of beauty in pyramids, that is not in spheres. And even more unique is a Galilean, or Celestial pyramid, which gives them special value. Galileo sees through the mannequin's bargain and realizes that if it became possible to make a thousand or a million copies of himself, while he would become virtually immortal, he would also become fungible: Galileo would be a commodity, like chicken breast.

The quote is from Philip Larkin's "Aubade" (the same as in "Nightfall I"). The mannequin is by Aurore Latuilerie, using the body of Bouguereau's *Young Shepherdess* at the San Diego Museum of Art and the head from the movie *I, Robot*. The portrait of Beethoven is by Carl Jäger. References are to the *Missa solemnis* and *Heiligenstadt Testament*. *The Temptation of Saint Anthony* by David Ryckaert III is at the Palazzo Pitti, Florence. The bust is Zeno of Citium (Herculaneum, Marble National Archeological Museum, Naples, courtesy of Livius. org). Zeno was known for his paradoxes (Achilles and the tortoise), with which he tried to show the incoherence of the idea of change. Here Zeno is faced with his own changes, death included. Zeno was a disciple of Parmenides, whom we met before (not only is there no change, but all is one). Indeed, mentor and pupil may have engaged in the traditional Greek practice of mixed intellectual and physical intercourse. Tymicha was a philosopher-mathematician herself. While we know that Zeno was tall and handsome, that she was short and ugly seems gratuitous slander. It is true, however, that they both challenged a tyrant in southern Italy. Zeno's end, crushed in a mortar, and his followers' response are reported by Diogenes Laertius. The mannequin may not be saying "perfect" and *"parfait"* in vain, as he may be referring to Derek Parfit (perfect), who has investigated the philosophical import of copying one's body and brain (*Star Trek* teleportation), with or without replacement (or branching, as he calls it). Courbet's *Self-Portrait with a Black Dog* is at the Petit-Palais, Paris.

The Terracotta Army was buried with the First Emperor of Qin (China) in 210–209 B.C., to help him rule another empire in the afterlife. By specific request of the emperor, no two individuals were to be alike (photograph by Robin Chen, Wikipedia). The collapsed cathedral is that of Ypres, Belgium, during World War I. The image of Europe at night is from NASA.

AFTERTHOUGHTS

33

Study Questions

"Judgment day has arrived," announced Alturi, standing in trinity between Frick and the bearded man.

Where am I? asked Galileo, wondering if he had returned to the classroom of his youth.

"Questions are the duty of life, and answers the privilege of Heaven," said Alturi sternly.

You know the answer to everything? ventured Galileo, still thinking of the mannequin and wondering if it might be there.

"Are errors never made in Paradise? Just try and do your best," replied Alturi with a smile.

The first question was from Frick: "You have found much consciousness in brains, or rather in certain parts of them. You sought to measure consciousness as integrated information, as Φ, its quality and its quantity a matter of numbers and geometry. You are a physicist, and thus I ask you: Physicist, what are your units?"

Galileo had contemplated this question. He knew that any kind of element or interval, in space and time, could have been chosen to measure Φ. What were the right elements? Neurons or groups of neurons, or areas of the brain, or entire brains, people, families, cities, nations, planets, or stars? Or smaller elements instead? Molecules or atoms or even smaller particles, of which the world was made? One could measure Φ at any of these levels and score a different number every time. And one could measure Φ over instants, seconds, or years and score a different number again. But which was the number of consciousness? If consciousness is a part of reality, it cannot vary depending on which unit is considered—one cannot be more or less conscious, here and now, depending on what one chooses to measure. I am an old man, said Galileo.

"There is no way out in Paradise," said Frick.

Then here is my answer, said Galileo. Φ can be measured at any scale in space and time, but in reality there will be a scale at which its maximum emerges. The scale at which integrated information condenses in space and time, the scales at which consciousness reaches its maximum, and thus comes into being—the scale at which the action happens.

"The scale at which the world is carved at its joints," said Frick.

Yes, said Galileo: Φ may be low for individual neurons, each one too feeble to effect much change by itself, though they be multitudes. And giant crowds of neurons are far too rough to make the fine distinctions that consciousness requires. But small groups of neurons will speak together loudly, be better heard by other groups, and together form a large and varied complex of surpassing Φ. There consciousness may indeed reside. The same in time: an instant is too short for neurons to converse—there is no time for hands to reach and shake: Φ for an instant will be zero. Likewise, Φ will grow tired after a day—a day is far too long for neurons—no matter what a group of neurons signaled in the morning, by night the results will be the same—all brain asleep. It is over the middle course, less than a second, that different signals from each group can make most difference to the rest, and Φ will reach its peak. And that, lo and behold, is the time scale at which consciousness flows.

Frick did not respond, so Galileo went on to ask: Tell me, in a brain made of neurons a thousand times faster than ours, will consciousness flow a thousand times faster? Does consciousness define its time and space, all from the inside? Define its elements, establish their identity over time, if their relationships stay the same?

"Paradise is a place for answers, not for questions," said Frick.

It was Alturi who asked the second question: "Is it not strange that consciousness isn't merely what is, but what could be?"

It is, said Galileo. Nothing is more immediately given than being conscious; consciousness is the most actual thing there is, yet it exists as a potential. It's strange indeed: I am conscious here and now, not only because my neurons are active or inactive in a certain way, but also by the countless states they could have occupied, and were ruled out by mechanisms. The experiences we shall never have must have been possible to experience, here and now, anything at all. Perhaps what Bruno said is true—*sostanza è possanza*—though it is strange.

"Perhaps it isn't strange for the thing itself," replied Alturi. "At each moment of time, what could have been is intrinsic to what is."

Just as mass is intrinsic to a body, thought Galileo. Once he himself had said that mass is what makes acceleration proportional to force. The more mass a body has, the less its speed changes when it receives a force. Yet it doesn't take a force to have a mass, he thought: a body doesn't need to calculate how its speed would change subject to a force, to actually have mass; nor does a brain need to compute its freedom—its repertoire in response to all perturbations, to actually experience it. So a brain's evanescent and elusive consciousness is as much material, or as little, as a body's mass, its most material aspect. Both are potential, yet both are actual. And just as mass can attract other bodies from without, so consciousness will enlighten the universe from within.

"The potential becomes actual: implex, simplex, complex," said Alturi. "All you could write when you face the blank page, all you could play when you touch the keyboard, all you could think when you close your eyes—is that the stuff experience is made of?"

Being before describing, said Galileo, and asked: Then what is integrated information, a fundamental property of the world?

"As fundamental as matter, perhaps more," said Alturi, "and perhaps the same thing—it from bit. For Φ is the fountain of phenomena. But now show me that you have learned your lessons, and tell me what exists."

I will tell you what I think I have learned, said Galileo. I have learned that information only exists if there is a difference that makes a difference. For how would anything exist, if nothing can make a difference to it? How could it exist, if it has no choice—an element with only one possible state? I have learned that information and causation are one and the same thing, and that is all there is: what exists must be a difference that makes a difference, a choice that's causal.

"Yes," said Alturi: "In the beginning is a gate both physical and logical, one taking two and choosing one, or taking one, where one is a special case of two. But does only one kind of mechanism exist, as said Anaximander, a NOR gate effecting space and time and consciousness? The simplest way to integrate two into one? Or does plurality reside at the core of the real, as said Empedocles, and there are AND and OR and NOT and COPY, and maybe more?"

Without understanding what Alturi meant, Galileo went on: I have also learned that information only exists if it is integrated—that only consciousness, integrated information—is *really* real, the only thing that exists *in and of itself,* and does not need another being to be. But simple aggregates do not exist in and of themselves—if you probe them, they dissolve into much lesser beings, a dust of dust. For you can group disparate objects in a thought, make something one that one is not—a heap of sand, a galaxy, or a crowd—but those exist as one just from the outside, from the perspective of a conscious being.

"Yes," said Alturi, "just like you exist and I exist, but not the two of us—*tertium no datur.*"

And finally, said Galileo, I have learned that integrated information only exists where it reaches a maximum—a maximum in elements, space, and time—and its existence excludes all the rest, for information cannot be counted twice.

"Yes," said Alturi. "Experience is definite and it has borders, borders that are cut with Occam's razor."

It was the old man's turn, and he said: "You learned that how consciousness is, its quality, is the shape of a quale, a constellation of bright points of integrated information. I ask you this: How is the world, then, in and of itself? How is the world in reality, if the world we see can only be the world we dream of?"

I learned that the geometry of the space of qualia contains all phenomena of the world, said Galileo, defining what blue is and what red is, what color and what form, what sight and sound, what image and what thought. I learned we only know the world as constructed by our brain, in its own image and likeness. That all is in the brain, not just color and taste, but also space and time, mass, number, and extension. That all is paradox: the brain, a small knot in the fabric of the world, is the sole source of the world itself. But then the world is the sole source of our brain. I learned we are the thing itself: at times sound and at times tragic, interminable or cruelly short, the only thing we experience from within. And I have learned that our selves should be large enough, that we can house our friends.

"The brain itself is part of the great world," said the old man. "The brain is built its way according to its laws, which rule the integration and the differentiation of its parts; due to the world it encountered, evolving through the seasons of the species, growing through its own development, learning through its own experience. The world is known not as reflected through a mirror. No, but by the shape to which the brain was hewed by the harsh hatchet of life—until its shape was such that its internal mechanisms could run in harmony with those external. Every perception is an act of memory, memory of the law, memory of history, and memory of experience."

I learned, said Galileo, that knowledge is the adjustment of inner to outer relations—a matching of unequals. That when we face the world, the more we know and understand, the more Φ grows and breathes, inner relations flourish, when they match outer ones. Beyond we cannot know, but guess, sometimes, that we were wrong.

"Henceforth be blind, for thou hast seen too much, and speak the truth that no man may believe," said the old man. "Consciousness is maximally irreducible, and it is unique; consciousness is the shape of understanding, the only shape that's really real—the most real thing there is; consciousness is a model of how the world might be, a model orig-

inal and proud, built over millions of years, with sacrifice immense, at the expense of innumerable losses, born on the life and death of countless ancestors, paid for by endless wars, won by the sweat of slaves, mended by wearing lessons, polished by civilizations, schools, and learned arguments, but it's a model woven in the only language we own, that of the brain and its enchanted loom. To know how much the world is in the image and likeness of our brain, it is enough to carry out your first experiment, to fall asleep and dream. What we dream of is what we know, what we can know is that of which we dream. The world can be imagined within, but not seen naked from without. And yet without the inner glow of consciousness, there would be no sight. What we must do is go and seek the light, the light that unifies."

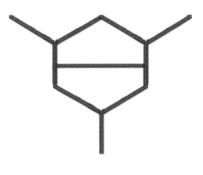

NOTES

Sostanza è possanza, substance is potential, was said by Giordano Bruno in his *De la causa, principio, et Uno* (1584). *Implex* is a term coined by Paul Valéry implying the great capacity of the mind in the space of possibilities. "It from bit" was John Archibald Wheeler's way of saying that all physical entities arise from underlying information content—or maybe from qubits (quantum bits). The quote is from Athena, caught in her nakedness in Tennyson's "Tiresias." The first figure is *The Three Wise Men* by Giorgione (at the Kunsthistorisches Museum, Vienna). The second is the interior of the Pantheon in Rome (a metaphor of the intrinsic perspective?) by Giovanni Paolo Panini. The third figure illustrates how much God's mantle in Michelangelo's

Sistine Chapel resembles the brain (Michelangelo had practiced dis-sections at length); the resemblance was pointed out in detail by F. L. Meshberger (*Journal of the American Medical Association,* 1990). As also shown by Emily Dickinson, artists can at times intuit more than they intend, or intend more than they intuit. So the giant synapse between God's and Adam's fingers signifies that the universe is created by the brain. The symbol at the end must hold some great significance for the author, but could not be deciphered.

ACKNOWLEDGMENTS

Over years of work on the problem of consciousness, I have enjoyed the support of many and likely taxed the patience of some. Chiara Cirelli and Lice Ghilardi have been invaluable to the conception and writing of *PHI* throughout its long gestation. My friend and colleague Christof Koch has assisted *PHI*'s delivery by promoting its publication, discussing its ideas with his unquenchable intellectual energy, and making essential contributions to the science of consciousness for more than two decades. I am especially indebted to Marcello Massimini, Olaf Sporns, and Randy McIntosh, whose help in growing the concepts presented in *PHI* has been priceless. I am also grateful to many with whom I have often discussed the nature of consciousness and from whose work I have drawn both knowledge and inspiration, among them Chris Adami, Mike Alkire, Giorgio Ascoli, Bernie Baars, David Balduzzi, Susan Blackmore, Ned Block, Mélanie Boly, Gyuri Buzsaki, David Chalmers, Patricia Churchland, the late Francis Crick, Jim Crutchfield, Antonio Damasio, Richie Davidson, Stan Dehaene, Dan Dennett, Gerald Edelman, Karl Friston, Stefano Fusi, Michael Gazzaniga, Allan Hobson, Tony Hudetz, Todd Hylton, Sid Kouider, Hakwan Lau, Steven Laureys, David Leopold, Rodolfo Llinas, Thomas Metzinger, Read Montague, Jim Olds, Brad Postle, Marc Raichle, Niko Schiff, Andy Schwartz, John

Searle, Wolf Singer, Nao Tsuchiya, and Barry van Veen. Despite burdening several friends with premature drafts of *PHI,* I have always received helpful advice. My thanks especially to Mike Alkire, Giorgio Ascoli, Ruth Benca, Gabriele Biella, Alex Borbèly, Marco Casonato, Paola d'Ascanio, Giovanna De Lorenzi, Ugo Faraguna, Fabio Ferrarelli, Stefano Fusi, Bob Golden, Sean Hill, Allan Hobson, Charlie Kaufman, Ned Kalin, Bill Linton, Cristiano Meossi, George Miklos, Yuval Nir, Pietro Pietrini, Brad Postle, Gian Luigi Salaris, Ernestina Schipani, Bob Shaye, Olaf Sporns, Irene Tobler, Eve van Cauter, and Vlad Vyazovskiy. Being constitutionally impaired at keeping track of correspondence, drafts, figures, and permissions, I had the good fortune of being assisted by Martha Pfister-Genskow, who lately managed to rescue the enterprise at the cost of much anxiety on her part. My father took upon himself the translation of an early draft of *PHI* into Italian and with the help of my mother imparted a certain urgency to my task—a small additional reason I owe them gratitude. When I began writing *PHI,* I had not really thought of an audience, let alone a publisher, and due to its unusual format it seemed unlikely that one would be interested. I was mistaken: Dan Frank at Pantheon examined *PHI* with a remarkably open mind and immediately offered his perceptive advice. Since then, working with him has been a true pleasure. One can afford the luxury of writing such a book only with generous support for the science that sustains it. I am grateful to the McDonnell Foundation, DARPA, the NIH Director's Pioneer Award, the NIMH and NINDS, and David White at Respironics. Finally, my heartfelt thanks go to the support provided by my laboratory, my department, the School of Medicine, the UW Medical Foundation, and the University of Wisconsin.

ILLUSTRATION CREDITS

18 Francis Crick in 1993, © 1993, Nick Sinclair/ Photo Researchers, Inc. © 2011 Photo Researchers, Inc. Image cropped by author.

19 *M82: Galaxy with a Supergalactic Wind,* Astronomy Picture of the Day, April 25, 2006. NASA, ESA, The Hubble Heritage Team (STScI/AURA). Acknowledgment: J. Gallagher (University of Wisconsin), M. Mountain (STScI), and P. Puxley (NSF).

20 Left image: *Mona Lisa,* ca. 1503–6 (oil on panel) (detail of 3179) by Leonardo da Vinci (1452–1519), Louvre, Paris, France / Giraudon / The Bridgeman Art Library. Image cropped by author. Right image: Two eyes of scallop (*Pecten*) plate 1e, pp. 116–17, from *Animal Eyes* by M. F. Land and D.-E. Nilsson (Oxford Animal Biology Series, © Oxford University Press, 2002). By permission of Oxford University Press (www.oup.com).

21 Hippocampal formation. Courtesy of Gyuri Buzsaki, MD, PhD, Rutgers University.

25 The Cloister (photo), French School (twelfth century) / Fontenay Abbey, Montbard, Burgundy, France / Bildarchiv Steffens / The Bridgeman Art Library.

26 Portrait of Nicolaus Copernicus (1473–1543) (oil on canvas) by Pomeranian School (sixteenth century), Nicolaus Copernicus Museum, Frombork, Poland / Giraudon / The Bridgeman Art Library. Image cropped by author.

26 *Crucifixion,* 1426 (tempera on panel), by Tommaso Masaccio (1401–1428), Museo e Gallerie Nazionali di Capodimonte, Naples, Italy / Giraudon / The Bridgeman Art Library. Image cropped by author.

27 Author image.

28 János Szentágothai, The Ferrier Lecture, 1977: *The Neuron Network of the Cerebral Cortex: A Functional Interpretation.* Proc. R. Soc. Lond. B. 201, No. 1144, 219–248 (1978), Figure 13 (p. 238). Permission granted by The Royal Society.

29 János Szentágothai, The Ferrier Lecture, 1977: *The Neuron Network of the Cerebral Cortex: A Functional Interpretation.* Proc. R. Soc. Lond. B. 201, No. 1144, 219–248 (1978), Figure 3 (p. 224). Permission granted by The Royal Society.

30 Nicholas D. Schiff, Urs Ribary, Diana Rodriguez Moreno, et al., "Residual Cerebral Activity and Behavioural Fragments in the Persistently Vegetative Brain." *Brain* 125 (June 2002) (Pt. 6): 1210–34, figure 4C, p. 1223. © 2002, Oxford University Press. Permission granted by Oxford University Press.

32 *Mary Magdalene,* ca. 1455 (polychrome and gilded wood) by Donatello (c. 1386–1466), Museo dell'Opera del Duomo, Florence, Italy/ The Bridgeman Art Library. Modification of image by author.

33 *Apollo and Daphne,* 1622–25 (marble), by Giovanni Lorenzo Bernini (1598–1680), Galleria Borghese, Rome, Italy/ Giraudon/ The Bridgeman Art Library. Modification of image by author.

37 *A Dance to the Music of Time,* ca. 1634–36 (oil on canvas) by Nicolas Poussin (1594–1665), © Wallace Collection, London, UK / The Bridgeman Art Library.

38 Author image modified from unidentified source.

39 *Portrait of the Artist,* 1650 (oil on canvas), by Nicolas Poussin (1594–1665), Louvre, Paris, France / Giraudon / The Bridgeman Art Library. Image cropped by author.

41 *Arcadian Shepherds* (oil on canvas) by Nicolas Poussin (1594–1665), Louvre, Paris, France / Giraudon / The Bridgeman Art Library.

42 Beni Isguen, © George Steinmetz. Permission to reproduce by George Steinmetz.

47 *The Allegory of Sight,* 1617 (oil on panel), by Jan Brueghel (1601–1678) and P. P.

Rubens (1577–1640), Prado, Madrid, Spain / The Bridgeman Art Library. Modification of image by author.

48 *Self-portrait* (oil on canvas) by Giovanni Paolo Lomazzo (1538–1600), Pinacoteca di Brera, Milan, Italy / The Bridgeman Art Library. Image cropped by author.

50 *The Fall of the Rebel Angels* (oil on panel) by Peter Paul Rubens (1577–1640), Alte Pinakothek, Munich, Germany / The Bridgeman Art Library. Modification of image by author.

51 *The Lady and the Unicorn: "Sight"* (tapestry) by French School (fifteenth century), Musée National du Moyen Age et des Thermes de Cluny, Paris / The Bridgeman Art Library. Image cropped by author.

54 *Self-portrait, 1556,* by Sofonisba Anguissola (ca. 1532–1625), Muzeum Zamek, Lancut, Poland / The Bridgeman Art Library. Image cropped by author.

58 Calculating machine, invented by Blaise Pascal (1623–1662) in 1642 (wood and metal) by French School (seventeenth century). CNAM, Conservatoire National des Arts et Métiers, Paris / Archives Charmet / The Bridgeman Art Library. Image cropped by author.

59 Church tower at Pennant Melangell, October 16, 2009, Powys, Wales (photograph). Credit: Gerald Morgan. Wikimedia Commons. Modification of image by author.

61 *The Game of Chess, 1555,* by Sofonisba Anguissola (ca. 1532–1625), Museum Narodowe, Poznan, Poland / The Bridgeman Art Library. Image modified by author by the addition of the Shadow Robot Hand developed by the Shadow Robot Company in London, © Shadow Robot Company 2008. Permission granted.

63 *Cardinal Richelieu on His Deathbed* (oil on canvas) by Philippe de Champaigne (1602–1674), Institut de France, Paris, France / Giraudon / The Bridgeman Art Library. Modification of image by author.

65 Colored galaxy is actually a visualization of millions of primes, © Adrian J. F. Leatherland http://www.mysteriousnumbers.com. Image kindly provided by Adrian J. F. Leatherland.

66 *The Nightmare, 1781* (oil on canvas), by Henry Fuseli (Fussli, Johann Heinrich) (1741–1825), Detroit Institute of Arts, USA / Founders Society purchase with Mr. and Mrs. Bert L. Smokler and Mr. and Mrs. Lawrence A. Fleischman funds / The Bridgeman Art Library.

70 *Portrait of a Lady* (oil on canvas) by Jacopo Robusti Tintoretto (1518–1594) (after) Worcester Art Museum, Massachusetts, USA / The Bridgeman Art Library.

72 Brainbow mouse. The brainbow mouse was produced by J. Livet, T.A. Weissman, H. Kang, et al. *Nature* 450 (2007): 56–62. © Tamily Weissman. Image kindly provided by Tamily Weissman.

73 Giorgio A. Ascoli, "Progress and Perspectives in Computational Neuroanatomy." *The Anatomical Record* (NEW ANAT) 257 (1999): 195–207, figure 6 (panel H), p. 204. Rat hippocampal slice © 1999 Wiley-Liss, Inc. and John Wiley and Sons, Inc. Permission granted by John Wiley and Sons, Inc.

74 Picture of neurons superimposed upon the hippocampal score. Courtesy of Gyuri Buzsaki, MD, PhD, Rutgers University.

75 *Barbara Strozzi* (oil on canvas) by Bernardo Strozzi (seventeenth century), Gemäldegalerie Alte Meister, Dresden, Germany / © Staatliche Kunstsammlungen Dresden / The Bridgeman Art Library.

78 Author image.

78 *Anatomical Machines, 1763–64,* by Giuseppe Salerno, Sansevero Chapel Museum,

Naples. Permission granted by Board of Directors of Museum Chapel Sansevero Ltd., Naples.

79 *Veiled Christ,* 1753 (marble), by Giuseppe Sanmartino, Sansevero Chapel Museum, Naples. Permission granted by Board of Directors of Museum Chapel Sansevero Ltd., Naples.

80 Achille-Louis Foville, *Traité complet* (1844), plate XV; artist E. Beau; engraver F. Bion.

81 *Modesty,* 1749–52 (marble), by Antonio Corradini, Sansevero Chapel Museum, Naples. Permission granted by Board of Directors of Museum Chapel Sansevero Ltd., Naples.

82 *Disillusion,* 1753–54 (marble), by Francesco Queirolo, Sansevero Chapel Museum, Naples. Permission granted by Board of Directors of Museum Chapel Sansevero Ltd., Naples.

83 Images combined of (left) *Disillusion,* 1753–54 (marble), by Francesco Queirolo, Sansevero Chapel Museum, Naples, and (right) *Modesty,* 1749–52 (marble), by Antonio Corradini, Sanservero Chapel Museum, Naples. Permission granted by Board of Directors of Museum Chapel Sansevero Ltd., Naples.

87 *The Ecstasy of Saint Margaret of Cortona,* by Giovanni Lanfranco (1582–1647), Palazzo Pitti, Florence, Italy / The Bridgeman Art Library. Image modified and cropped by author.

88 Passionate attitudes: Augustine in ecstasy, from "Iconographie de la Salpetriere" by Desire Magloire Bourneville (1840–1909) and Paul Regnard (1850–1927), 1878. Black-and-white photograph by Albert Londe (fl. 1878–89). Bibliotheque de la Faculte de Medecine, Paris, France / Archives Charmet / The Bridgeman Art Library. Image cropped by author.

89 *Abelard Soliciting the Hand of Héloise,* by Angelica Kauffmann (1741–1807), © Burghley House Collection, Lincolnshire, U.K. / The Bridgeman Art Library. Image cropped by author.

91 *A Clinical Lesson with Dr. Charcot at the Salpétrière,* 1887 (oil on canvas), black-and-white photograph by Pierre Andre Brouillet (1857–ca. 1920), Faculty of Medicine, Lyon, France / The Bridgeman Art Library. Image cropped by author.

92 *The Bird Organ; or, A Woman Varying Her Pleasures,* 1751 (oil on canvas), by Jean-Baptiste-Siméon Chardin (1699–1779), Louvre, Paris, France / Giraudon / The Bridgeman Art Library. Image cropped by author.

94 *The Ecstasy of Saint Teresa* (marble), by Giovanni Lorenzo Bernini (1598–1680), Santa Maria della Vittoria, Rome, Italy / The Bridgeman Art Library. Image cropped by author.

98 *Maddalena Svenuta,* 1663, by Guido Cagnacci, Galleria Nazionale d'Arte Antica, Palazzo Barberini, Rome. With the authorization of the Soprintendenza Speciale per il Patrimonio Storico, Artistico ed Etnoantropologico e per il Polo Museale della città di Roma.

99 *The Passion of Joan of Arc (La passion de Jeanne d'Arc,* France), 1928. Director Carl Dreyer. Credit: M. J. Gourland/Photofest.

100 *The Passion of Joan of Arc (La passion de Jeanne d'Arc,* France), 1928. Director Carl Dreyer. Credit: M. J. Gourland/Photofest.

101 Author image.

102 *The Passion of Joan of Arc (La passion de Jeanne d'Arc,* France), 1928. Director Carl Dreyer. Credit: M. J. Gourland/Photofest.

105 *Storm at Sea* (oil on panel), by Willem van de Velde the Younger (1633–1707), Worcester Art Museum, Massachusetts, USA / The Bridgeman Art Library. Image cropped by author.

107 *Philosopher in Meditation,* 1632 (oil on panel), by Rembrandt Harmensz van Rijn (1606–1669), Louvre, Paris, France / The Bridgeman Art Library.

108 Portrait of René Descartes (1596–1650), c. 1649 (oil on canvas), by Frans Hals (1582 or 1583–1666) (after), Louvre, Paris, France / Giraudon/ The Bridgeman Art Library. Image cropped by author.

110 *Crâne de Descartes (René Descartes's skull)*, Muséum National d'Histoire Naturelle, Musée de l'Homme, Paris. Credit: © M.N.H.N-Daniel Ponsard. Image kindly provided by the Muséum National d'Histoire Naturelle, Paris.

111 *The Incredulity of Saint Thomas,* 1602–3 (oil on canvas), by Michelangelo Merisi da Caravaggio (1571–1610), Schloss Sanssouci, Potsdam, Brandenburg, Germany / Alinari / The Bridgeman Art Library.

118 Alan Turing, King's College Library, Cambridge. AMT/K/7/13, © Turing family. Permission to reproduce by King's College, Cambridge, U.K., and the Turing family. Image cropped by author.

120 Jacquard's loom, showing the threads and the punched cards (Frenchman Joseph Marie Jacquard invented a power loom) found at http://www.computersciencelab.com/ComputerHistory/HistoryPt2.htm. Unknown source.

121 Author image.

130 *Galileo Galilei* (1564–1642) (oil on canvas), by Justus Sustermans (1597–1681), (school of) Galleria Palatina, Palazzo Pitti, Florence, Italy / The Bridgeman Art Library. Modification of image by author.

131 Emperor Yongzheng, Chinese School, Wikimedia Commons. Modification of image by author with the addition of Babbage's analytical engine, 1834, by Charles Babbage © Science Museum/SSPL.

136 4096 Color Wheel, Version 1.3. Credit Jemima Pereira.

137 *Portrait of a Woman at Her Toilet,* 1512–15 (oil on canvas), by Titian (Tiziano Vecellio) (ca. 1488–1576), Louvre, Paris, France / Giraudon / The Bridgeman Art Library.

138 Author image.

138 Author image.

138 Author image.

140 *Koyaanisqatsi* (1982). Director Godfrey Reggio. Credit: New Cinema/Photofest.

141 Author image.

142 Unicycle juggling act, ca. 1923 (black-and-white photograph), Library of Congress, Washington, D.C.

143 Claude Shannon with mouse (black-and-white photograph). Credit: Alcatel-Lucent / Bell Labs, © Alcatel-Lucent USA Inc. Reprinted with permission of Alcatel-Lucent USA Inc. Image cropped by author.

144 *The Most Beautiful Machine,* © 2003 Hanns-Martin Wagner (www.kugelbahn.ch). Image kindly provided by Hanns-Martin Wagner.

148 *Galileo Galilei* (1564–1642) (oil on canvas), by Justus Sustermans (1597–1681), (school of) Galleria Palatina, Palazzo Pitti, Florence, Italy / The Bridgeman Art Library. Modification of image by author.

149 Author image.

150 *The Head of an Old Man,* by Jacob Jordaens (1593–1678) / © The Trustees of the Weston Park Foundation, U.K. / The Bridgeman Art Library.

152 *Democritus; or, The Man with a Globe* (oil on canvas), by Diego Rodriguez de Silva y Velázquez (1599–1660), Musée des Beaux-Arts, Rouen, France / Giraudon / The Bridgeman Art Library. Image cropped and modified by author.

152 *Lucretius* (marble). Wikimedia Commons. Unknown source. Image cropped by author.

152 Portrait of Julien Offroy de La Mettrie (1709–1751) (engraving) (black-and-white photograph), by Johann Christian Gottfried Fritzsch (ca. 1720–1802). Private collection / Roger-Viollet, Paris / The Bridgeman Art Library. Image cropped and modified by author.

152 Portrait of René Descartes (1596–1650), ca. 1649 (oil on canvas), by Frans Hals (1582 or 1583–1666) (after), Louvre, Paris, France / Giraudon / The Bridgeman Art Library. Image cropped and modified by author.

154 *The School of Athens,* detail of a figure from the left-hand side, 1510–11 (fresco) (detail of 87597), by Raphael (Raffaello Sanzio of Urbino) (1483–1520), Vatican Museums and Galleries, Vatican City, Italy / The Bridgeman Art Library. Image cropped by author.

154 *Gerolamo Cardano.* Wikimedia Commons. Unknown source. Image cropped by author.

154 Portrait of René Descartes (1596–1650), ca. 1649 (oil on canvas), by Frans Hals (1582 or 1583–1666) (after), Louvre, Paris, France / Giraudon / The Bridgeman Art Library. Image cropped by author.

154 *Immanuel Kant.* Wikimedia Commons. Image cropped by author.

157 William James. Notman Studio, photographer. Call number *2002M-44. By permission of the Houghton Library, Harvard University. Image cropped by author.

158 Left: the Northern Galactic Hemisphere (© Tunç Tezel). Right: The Southern Galactic Hemisphere (image © Stéphane Guisard).

159 *A Tale of Two Hemispheres,* Astronomy Picture of the Day, July 30, 2011 (http://apod.nasa.gov/apod/ap110730.html), © Stéphane Guisard and Tunç Tezel.

160 Author image.

161 Author image.

161 Bottom: Achille-Louis Foville, *Traité complet* (1844), plate XV; artist E. Beau; engraver F. Bion.

162 Author image.

164 Author image.

165 Author image.

165 Author image.

166 Author image.

167 Author image.

170 *The Ripe Harvest,* 1924 (pen, watercolor, and pencil on paper on cardboard), by Paul Klee (1879–1940), Sprengel Museum, Hanover, Germany / The Bridgeman Art Library. Image cropped by author.

171 *Portrait of Adam* (oil) by Giuseppe Arcimboldo (1527–1593), private collection, Switzerland / The Bridgeman Art Library. Modification of image by author.

173 *Plato's Cave,* 1998, © Kenneth Eward. Image kindly provided by Kenneth Eward.

175 *The Boy Exposing a Bat to the Flame* (oil on canvas), by Trophime Bigot (c. 1595–1650), Galleria Doria Pamphilj, Rome, Italy / Alinari / The Bridgeman Art Library.

176 *The Bat,* 1522 (watercolor on paper), by Albrecht Dürer or Duerer (1471–1528), Musée des Beaux-Arts et d'Archeologie, Besancon, France / Giraudon / The Bridgeman Art Library. Image cropped by author.

178   K. P. Bhatnagar, "The Brain of the Common Vampire Bat, *Desmodus rotundus muri-nus* (Wagner, 1840): A Cytoarchitectural Atlas." *Brazilian Journal of Biology.* Figure 1(a): Hemisected head of female Desmodus rotundus showing the brain in its major subdivisions. Permission to reproduce by the *Brazilian Journal of Biology* 68 no. 3 (2008): 583–99.

179   A green moray eel (*Gymnothorax prasinus*) and greyface moray eel (*Gymnothorax thyrsoideus*) sharing a lair. Green Island, South West Rocks, New South Wales, © 2006 Richard Ling, www.rling.com. Wikimedia Commons. Image cropped by author.

183   *The Astronomer,* 1668 (oil on canvas), by Jan Vermeer (1632–1675), Louvre, Paris, France / Giraudon / The Bridgeman Art Library. Modification of image by author.

184   Human connectome: network of connections in the brain. Courtesy of Olaf Sporns, PhD, Indiana University, and Patric Hagmann, PhD, EPFL Switzerland.

186   The Imperial Gallery of the basilica (photo), Haghia Sophia, Istanbul, Turkey / The Bridgeman Art Library.

187   *Melancolia I,* 1514, by Albrecht Dürer. Library of Congress, Washington, D.C.

188   Ugolino gnawing the head of Ruggieri, by Gustave Doré. Illustrating Canto XXXII of *The Divine Comedy, Inferno,* by Dante Alighieri. Author image from a print.

190   Brain eroded from Alzheimer's disease, courtesy of Dr. Paul Thompson of the Laboratory of Neuro Imaging at the University of California, Los Angeles.

191   *Chicago Board of Trade II,* 1999, by Andreas Gursky, © 2011 Andreas Gursky / Artists Rights Society (ARS), New York / VG Bild-Kunst, Bonn. Permisssion to reproduce by Artists Rights Society (ARS). Image kindly provided by the artist, Andreas Gursky, courtesy of Sprüth Magers, Berlin, London.

192   30: Flawless pear-shaped rose-colored diamond of 16.10 cts., private collection. Photo © Christie's Images / The Bridgeman Art Library. Image cropped and modified by author.

195   *Carceri VII,* "The Drawbridge," 1760 (etching), by Giovanni Battista Piranesi (1720–1778). On loan to the Hamburg Kunsthalle, Hamburg, Germany / The Bridgeman Art Library.

196   Portrait of Gottfried Wilhelm Leibniz (1646–1716) (oil on canvas), by German School (eighteenth century). Niedersachsisches Landesmuseum, Hanover, Germany / Flammarion / The Bridgeman Art Library. Image cropped by author.

197   *Stepped Reckoner,* 1904, by Gottfried Wilhelm Leibniz, in Hannover. Image credit: Wilhelm Franz Meyer. Wikimedia Commons.

198   *Woman with a Candle* (oil on canvas), by Godfried Schalken or Schalcken (1643–1706), Palazzo Pitti, Florence, Italy / The Bridgeman Art Library. Image cropped by author.

200   "Biological Object Recognition," *Scholarpedia,* adapted from T. Serre, M. Kouh, C. Cadieu, et al., *A Theory of Object Recognition: Computations and Circuits in the Feedforward Path of the Ventral Stream in Primate Visual Cortex.* In Artificial Intelligence Memo #2005–2036. Boston: MIT, 2005. Permission to reproduce by Gabriel Kreiman, PhD, Harvard.

202   Author image based on *Ancient Harmony,* 1925 (no. 236) (oil on cardboard), by Paul Klee (1879–1940), Kunstmuseum, Basel, Switzerland / Gift of Richard Doetsch-Benziger, 1960 / The Bridgeman Art Library.

205   Author image.

206   Author image.

207 Author image.

210 *Haze.* Courtesy of Nancy Lobaugh of Toronto, Canada.

211 Author image.

213 *Rosa Celeste,* nineteenth century, by Gustave Doré from Dante Alighieri, "Canto XXXI," *The Divine Comedy by Dante.* Author image from a print.

214 Image created using Robert Webb's Stella4D software, available from http://www .software3d.com/Stella.php, © Robert Webb. Image kindly provided by Robert Webb.

216 Kepler's Platonic solid model of the solar system from *Mysterium cosmograhicum* (1596), by Johannes Kepler (1571–1630). Wikimedia Commons.

219 *NGC 253: Dusty Island Universe,* Astronomy Picture of the Day, November 21, 2009, © Star Shadows Remote Observatory (SSRO) and PROMPT/CTIO (Steve Mazlin, Jack Harvey, Rick Gilbert, and Daniel Verschatse). Permission granted by SSRO. Image kindly provided by SSRO.

220 *Caroline Herschel (1750–1848),* 1829, by Martin Francois Tielemans (1784–1864). Private collection / The Bridgeman Art Library. Image cropped by author.

222 *Threads of Light,* © Kristian Cvecek, http://quit007.deviantart.com/. Image kindly provided by Kristian Cvecek.

223 Scene of owl perched in tree silhouetted against full moon low to the horizon. Credit: Kirk Woellert/National Science Foundation (NSF).

224 Comet Hale-Bopp above a tree, March 29, 1997, © Philipp Salzgeber, http:// salzgeber.at. Permission granted by Philipp Salzgeber.

225 *Ulysses and Diomed Swathed in the Same Flame,* 1824 (watercolor over pencil on paper), by William Blake (1757–1827), National Gallery of Victoria, Melbourne, Australia / The Bridgeman Art Library.

225 *The Path of the Sun Through the Stars on the Night of the 4th July 1442,* from the soffit above the altar, ca. 1430 (fresco), by Giuliano d'Arrighi Pesello (1367–1446), Old Sacristy, San Lorenzo, Florence, Italy / The Bridgeman Art Library. Image cropped by author.

229 Author image.

230 *Mary Magdalene with a Night Light,* 1630–35 (oil on canvas), by Georges de La Tour (1593–1652), Louvre, Paris, France / Giraudon / The Bridgeman Art Library.

231 *Charles Darwin (1809–1882)* (photograph), by Julia Margaret Cameron (1815–1879). Private collection / The Stapleton Collection / The Bridgeman Art Library.

234 *Giordano Bruno,* 1889 (bronze statue), by Ettore Ferrari (1845–1929), Campo dei Fiori, Rome. Credit: Jastrow. Wikimedia Commons.

235 Tomb of Ilaria del Carretto, ca. 1406–8 (marble), by Jacopo della Quercia, Lucca Cathedral, Lucca. The Project Gutenberg eBook, *Tuscan Sculpture of the Fifteenth Century,* by Estelle M. Hurll.

236 The Vietnam Veterans Memorial with the National Monument—Washington, D.C., © Glyn Lowe Photos. Image cropped by author.

236 *Night and Fog (Nuit et brouillard),* 1955. Director Alain Resnais. Credit: Argos Films/ Photofest.

237 *The Massacre of the Innocents,* 1301 (marble), by Giovanni Pisano, Sant'Andrea church, Pistoia. Author image.

238 *Resurrection of Lazarus* (oil on canvas) (also see 232098,94,95), by Michelangelo Merisi da Caravaggio (1571–1610), Museo Regionale, Messina, Sicily, Italy / Giraudon / The Bridgeman Art Library. Image cropped by author.

240 *Rondanini Pietà,* detail of the heads of Christ and Mary (marble), by Michelangelo
Buonarroti (1475–1564), Castello Sforzesco, Milan, Italy / Giraudon / The Bridgeman Art library. Image cropped by author.

243 Author image modified from unidentified source.

245 *The Tower of Babel,* 1563 (oil on panel) (for details see 93768–69, 186437–186438), by
Pieter Brueghel the Elder (ca. 1525–1569), Kunsthistorisches Museum, Vienna, Austria / The Bridgeman Art Library.

246 Author image.

247 Author image.

248 *Saint Robert Bellarmine.* Unknown source, Wikimedia Commons.

249 *Self-Portrait,* 1996 (mixed media on paper), by William Utermohlen (1933–2007).
Private collection / The Bridgeman Art Library.

250 *Oval Head,* 2002 (pencil on paper), by William Utermohlen (1933–2007). Private collection / The Bridgeman Art Library.

253 Babbage's Analytical Engine, 1834, by Charles Babbage. © Science Museum/SSPL.
Permission to reproduce by Science Museum/SSPL. Image cropped by author.

254 Wilder Penfield and Edwin Boldrey, "Somatic Motor and Sensory Representa-
tion in the Cerebral Cortex of Man as Studied by Electrical Stimulation." *Brain* 60,
no. 4 (1937): 389–443, figure 21. © 1937, Oxford University Press, by Permission of
Oxford University Press.

257 *Kopf 25–03–1989* (canvas, 198.6 cm × 250.4 cm) by Armando. Credit: © Dordrecht,
Dordrechts Museum. Permission granted by the artist, Armando, and image kindly
provided by Dordrechts Museum.

258 Author image.

263 *Landscape with the Fall of Icarus,* ca. 1555 (oil on canvas), by Pieter Brueghel the Elder
(ca. 1525–1569), Musées Royaux des Beaux-Arts de Belgique, Brussels, Belgium /
Giraudon / The Bridgeman Art Library.

264 *Izhevsk: the remnants of the old order still remain,* © Paul Artus http://artus.orconhosting
.net.nz/. Modification of image by author.

265 *Zeus and Io,* by Correggio (Antonio Allegri) (ca. 1489–1534), Kunsthistorisches
Museum, Vienna, Austria / The Bridgeman Art Library.

266 *Earth; or, The Earthly Paradise,* detail of a cow, porcupine, and other animals, 1607–8
(oil on copper) (detail of 93899), by Jan Brueghel the Elder (1568–1625), Louvre,
Paris, France / Peter Willi / The Bridgeman Art Library.

267 *Narcissus,* ca. 1597–99 (oil on canvas), by Michelangelo Merisi da Caravaggio (1571–
1610), Palazzo Barberini, Rome, Italy / The Bridgeman Art Library. Modification
of image by author.

270 *Esel* (date unknown), by Johann Georg Grimm (1846–1887), Bühl am Alpsee,
Bavaria. Image kindly provided by the Grimm Vereinigung, Immenstadt.

271 *Balthazar and Marie (Au Hasard Balthazar),* 1966. Director Robert Bresson, © Mme
Bresson. Permission granted by Mme Mylène Bresson.

272 *Asino (Donkey),* 1956 (oil on canvas), by Tino Vaglieri (1929–), Museo Civico Bodini,
Floriano Bodini Civic Museum (Gemonio), property of Paola e Sara Bodini. Permission granted by Museum Floriano Bodini (Gemonio).

273 *Capriccio* scene: Animals in the sky (oil on canvas) by Francisco José de Goya y Lucientes (1746–1828), Musée des Beaux-Arts, Agen, France / Giraudon / The Bridgeman Art Library.

274 Author image.

277 *Pregnant Woman and Death,* 1911, by Egon Schiele (1890–1918), Narodni Galerie, Prague, Czech Republic / The Bridgeman Art Library.

278 *Quaestiones medico-legales,* 1657, by Paolo Zacchia. National Library of Medicine, NIH.

280 K.-K. Cheung, S. C. Mok, P. Rezaie, and W.Y. Chan, "Dynamic Expression of Dab2 in the Mouse Embryonic Central Nervous System," *BMC Developmental Biology* 8 (2008): 76, fig. 1, panel C' (lateral view), E9.5, Dab Immunoreactivity in the Neural Tube. © 2008 Cheung et al.; licensee BioMed Central Ltd., BioMed Central open access license agreement.

282 *The Carcass of an Ox,* late 1630s (oil on panel), by Rembrandt Harmensz van Rijn (1606–1669), Art Gallery and Museum, Kelvingrove, Glasgow, Scotland / © Culture and Sport Glasgow (Museums) / The Bridgeman Art Library.

283 *Saturn Devouring One of His Children,* 1821–23 (oil on canvas), by Francisco José de Goya y Lucientes (1746–1828), Prado, Madrid, Spain / The Bridgeman Art Library.

285 *The City of Lost Children (La cité des enfants Perdus),* 1995. Director Marc Caro, Jean-Pierre Jeunet. Credit: Sony Pictures Classics/Photofest.

286 Eye stairwell from the ground, © 2005 Cindy Mosqueda.

287 Drawing of a flea from *Micrographia,* 1665, by Robert Hooke. Wikimedia Commons.

288 *The Agony of "The Endurance,"* from *Expedition to the South Pole* by Ernest Shackleton (1874–1922), 1914–17 (black-and-white photograph) by French photographer (twentieth century), Fondation Paul-Emile Victor, Paris, France / Archives Charmet / The Bridgeman Art Library. Image cropped by author.

290 *The Wanderer Above the Sea of Fog,* 1818 (oil on canvas), by Caspar David Friedrich (1774–1840), Hamburger Kunsthalle, Hamburg, Germany / The Bridgeman Art Library.

291 Author image.

295 Emily Dickinson, ca. 1848. Photographic print of a daguerreotype (scratched original). Todd-Bingham Picture Coll Manuscripts and Archives, Yale University.

298 *The Tower of Babel,* 1679 (engraving) (black-and-white photograph), by Athanasius Kircher, (1602–1680), private collection / The Bridgeman Art Library. Image cropped by author.

299 *Portrait of Don Luis de Góngora y Argote* (1561–1627), 1622 (oil on canvas), by Diego Rodriguez de Silva y Velázquez (1599–1660), Prado, Madrid, Spain / Giraudon / The Bridgeman Art Library.

300 Left: *Pollock Tribute No. 4,* by Barry Tickle, http://movingpaintings.co.uk. Image kindly provided by Barry Tickle. Right: Gogi-Cox stain of cerebral rat neurons, author image.

301 *Self-Portrait at the Mirror,* by Parmigianino (Francesco Mazzola) (1503–1540), Kunsthistorisches Museum, Vienna, Austria / Ali Meyer / The Bridgeman Art Library. Image cropped by author.

302 *8½,* 1963. Director Federico Fellini. Credit: Embassy Pictures Corporation/Photofest.

303 Bust of Costanza Buonarelli (marble), by Giovanni Lorenzo Bernini (1598–1680), Museo Nazionale del Bargello, Florence, Italy / The Bridgeman Art Library.

304 *The Rape of Proserpina* (marble), by Giovanni Lorenzo Bernini (1598–1680), Galleria Borghese, Rome, Italy / Alinari / The Bridgeman Art Library. Image cropped by author.

306 *King of Quails* (watercolor), n. 1988 ORN, by Jacopa Ligozzi (1547–1632) / Gabinetto

dei Disegni e delle Stampe, Uffizi, Florence, Italy. Image © Scala / Art Resource, N.Y.

307 The Vatican, Rome (engraving), by Giovanni Battista Falda (ca. 1648–1678), private collection / The Bridgeman Art Library. Image cropped by author.

308 *The "Atlas" Slave,* ca. 1519–23 (marble), by Michelangelo Buonarroti (1475–1564), Galleria dell' Accademia, Florence, Italy / Alinari / The Bridgeman Art Library. Image cropped by author.

311 *The Country School,* by Winslow Homer (1836–1910), St. Louis Art Museum, Missouri, USA / The Bridgeman Art Library.

314 *The Cuman Sibyl,* from the Villa Carducci series of famous men and women, ca. 1450 (fresco), by Andrea del Castagno (1423–1457), Galleria degli Uffizi, Florence, Italy / The Bridgeman Art Library. Modification of image by author by the addition of Shadow Dexterous Robot Hand holding a lightbulb, 2008, developed by the Shadow Robot Company in London, © 2008 Shadow Robot Company. Permission granted.

315 *My Friend Ernest (Paris, 1929)* (silver gelatin print), by André Kertész (1894–1985), The Israel Museum, Jerusalem, Israel / The Noel and Harriette Levine Collection / The Bridgeman Art Library. Copyrights for Kertész André (dit), Kertész Andor (1894–1985): © RMN—Gestion droits d'auteur, © Collection Centre Pompidou, Dist. RMN.

317 *Sam and the Perfect World,* 2005 (oil on linen), by David Lenz. Milwaukee Art Museum. Image © David Lenz.

318 *Eternity* (marble), by Pietro Daverio. Allegorical caryatid from the Monument to Charles Borromeo in the apse of the Milan Cathedral (1611). Credit: © Giovanni Dall'Orto, July 14, 2007. Wikimedia Commons. Image cropped by author.

323 *iBouguereau,* by Aurore Latuilerie. Modification of *The Young Shepherdess,* 1885 (oil on canvas mounted on board), by William-Adolphe Bouguereau (1825–1905), San Diego Museum of Art, USA / gift of Mr. and Mrs. Edwin S. Larsen / The Bridgeman Art Library. With the head from *I, Robot,* 2004. Director Alex Proyas. Credit: Twentieth Century Fox / Photofest.

325 *Ludwig van Beethoven,* by Carl Jäger. (1833–1887) Library of Congress, Washington, D.C.

327 *The Temptation of Saint Anthony of Egypt* by David Ryckaert III (1612–1661), Palazzo Pitti, Florence, Italy / The Bridgeman Art Library.

328 *Zeno of Citium* (marble), Herculaneum National Archaeological Museum, Naples. © Livius.org. Image kindly provided by Livius.org.

331 *Self-Portrait with a Black Dog,* 1842 (oil on canvas), by Gustave Courbet (1819–1877), Musée de la Ville de Paris, Musée du Petit-Palais, France / The Bridgeman Art Library.

334 Terracotta warriors, Xian Museum. Credit: Robin Chen. Wikimedia Commons.

335 Ypres, Belgium, 1919. Photographic print (gelatin silver). Credit: William Lester King, Library of Congress, Washington, D.C.

337 Europe at night. Credit: U.S. Defense Meteorological Satellites Program (DMSP), and NASA/Goddard Space Flight Center Scientific Visualization Studio. NASA.

337 Author image.

343 *The Three Wise Men,* by Giorgione (Giorgio da Castelfranco) (1476 or 1478–1510), Kunsthistorisches Museum, Vienna, Austria / The Bridgeman Art Library.

345 *The Interior of the Pantheon, Rome, Looking North from the Main Altar to the Entrance,*

1732 (oil on canvas), by Giovanni Paolo Pannini or Panini (1691 or 1692–1765), private collection/ photo © Christie's Images/ The Bridgeman Art Library. Image cropped by author.

347 Sistine Chapel ceiling (1508–12): *The Creation of Adam,* 1511–12 (fresco) (postrestoration), by Michelangelo Buonarroti (1475–1564), Vatican Museums and Galleries, Vatican City, Italy / The Bridgeman Art Library. Image modified and cropped by author.

349 Author image.

ABOUT THE AUTHOR

Giulio Tononi, a neuroscientist and psychiatrist, is best known for his integrated information theory of consciousness and for his work on another scientific riddle—the function of sleep. He is a professor of psychiatry, the David P. White Professor of Sleep Medicine, and the Distinguished Chair in Consciousness Science at the University of Wisconsin. In addition to the major scientific journals, his research has appeared in *New Scientist, Science Daily,* and *Scientific American* and has been featured in *The New York Times* and *The Economist.* He is the coauthor, with Nobel laureate Gerard Edelman, of *A Universe of Consciousness* (2000).

A NOTE ON THE TYPE

This book was set in a version of the well-known Monotype face
Bembo. This letter was cut for the celebrated Venetian printer
Aldus Manutius by Francesco Griffo, and first used in Pietro Cardi-
nal Bembo's *De Aetna* of 1495.

The companion italic is an adaptation of the chancery script type
designed by the calligrapher and printer Lodovico degli Arrighi.

Composed by North Market Street Graphics, Lancaster, Pennsylvania

Printed and bound by Tien Wah Press, Singapore

Designed by Maggie Hinders

31901051512285